T0127678

WAVES AND PARTICLES

Two Essays on Fundamental Physics

WAVES AND PARTICLES

Two Essays on Fundamental Physics

Roger G. Newton
Indiana University, USA

NEW JERSEY • LONDON • SINGAPORE • BEIJING • SHANGHAI • HONG KONG • TAIPEI • CHENNAI

Published by

World Scientific Publishing Co. Pte. Ltd.
5 Toh Tuck Link, Singapore 596224
USA office: 27 Warren Street, Suite 401-402, Hackensack, NJ 07601
UK office: 57 Shelton Street, Covent Garden, London WC2H 9HE

Library of Congress Cataloging-in-Publication Data
Newton, Roger G., author.
Waves and particles : two essays on fundamental physics / Roger G. Newton, Indiana University, USA.
pages cm
Includes bibliographical references and index.
ISBN 978-9814449670 (softcover : alk. paper)
1. Wave mechanics. 2. Particles (Nuclear physics) 3. Quantum theory. I. Title.
QC174.2.N49 2014
531'.1133--dc23
2013048542

British Library Cataloguing-in-Publication Data
A catalogue record for this book is available from the British Library.

Typeset by Stallion Press
Email: enquiries@stallionpress.com

Printed in Singapore

Contents

Prologue

Physics, or natural philosophy as it used to be called, is concerned with all the inanimate aspects of the world. While many of the questions it seeks to answer have been asked for millennia, and a variety of tentative answers have been offered in most civilizations, those that have turned out to make up what we now call science have withstood the most searching tests. There is, of course no guarantee that the so-called scientific method, the essence of which is about five centuries old, will always lead to results that remain accepted as valid forever, but most of them have been found reliable.

These essays are meant for readers who are not trained in rigorous sciences but have an interest in understanding the world, including many very familiar phenomena the nature of which is unknown to them. Among all the various fields of physics this little volume concentrates on two that may be called structure and dynamics, or parts of dynamics. While dynamics, of course, encompasses the large areas of forces, response to forces, and the motions of objects, there is a great variety of kinds of motions. While not claiming that waves are the most important kind of motion — indeed it is not even clear to start with that all waves are a kind of motion — I will single them out as of particular interest. There are many different sorts of waves, some of them not recognized as such by the uninformed.

The structure of the world has puzzled philosophers for millennia. Is all of matter made up of water, fire, and earth as many of the ancient Greeks believed, or is it made of particles called atoms? The idea of a world made up of particles, too, has arisen in many ancient cultures, though perhaps nowhere with as much specificity as in Greece. We shall be concerned not only with atoms but with a great many more recently found particles, discussing both the enormous new machines needed for their discovery and the mathematical tools required to make sense of them. Their variety was so great that new methods were needed to lead physicists to an understanding comparable to that furnished by Mendeleyev for chemists by his periodic table of the elements.

I hope that readers will not only learn aspects of science from this book, but that they will enjoy it.

Part I

Tsunami and other Waves

Introduction to Part I

We all remember the severe damage to a nuclear power plant in Tohoku, Japan, in 2011, as well as the great loss of life and the enduring radioactivity caused by an enormous tidal wave due to an earthquake. It engraved the word *tsunami* forever on our memory. Of course, it wasn't the first event of this nature; there have been many others before, equally or even more destructive.

In this little essay we shall describe tsunami[1] and then compare this wave to other kinds of waves and their properties and underlying laws. Rather than describing the detailed history of tsunami and their well-known causes, earthquakes and volcanic eruptions, our aim is eventually to be able to explain tsunami by means of mathematical equations just as other waves are explained. To set the stage, after a vivid description of a tsunami event in Japan in the nineteenth century, we briefly trace the long history of explaining phenomena by means of equations, from the ancient Greeks to the nineteenth century. Then, to begin our comparison of tsunami with more common kinds of waves, we describe those of sound. The production of musical harmonies by the vibrations of string instruments, which confirm some of the conjectures of the ancient philosopher Pythagoras, is translated into sound waves, governed by laws common

[1]The word being of Japanese origin, its plural is tsunami, just as the plural of fish is fish.

to other kinds of waves to be described. We shall discuss not only facts about the acoustics of concert halls, but also ultrasound and its variety of uses, as well as shock waves.

We then turn to electromagnetic waves, beginning with light waves and such phenomena as refraction and the fascinating heavenly ones of the blue sky, the rainbow and the glory. Very rarely observed, a picture of the glory will show that it is well worth seeing if you have the chance. Other electromagnetic waves employed for important purposes are microwaves and radio waves; these are used for radar, microwave ovens, and radio astronomy. We shall describe all of them, including the fundamental discovery about the universe made by two radio astronomers. This essay will also describe the basics of how a radio works.

The next waves described arise in the new kinds of physics introduced in the twentieth century: quantum mechanics and Einstein's general theory of relativity. Schrödinger's wave mechanics is fundamental to the quantum theory, and Einstein predicted the existence of gravitational waves, which are very difficult to detect.

Finally, there are, of course, water waves. Some of these we are familiar with, on lakes and oceans, including the tides, but there are also more unusual ones, first seen in the nineteenth century on shallow Dutch canals: solitary waves, later called solitons, that behave quite differently from all the other kinds of waves described. That brings us back to our starting point, the tsunami. Are tsunami waves and their enormous destructive power explicable in the same terms as other ocean waves, or are they more akin to solitary waves and their strange behavior seen on Dutch canals? The final answer to this question is still unknown and subject to ongoing research. No doubt the right set of equations to explain tsunami will eventually be found.

1

The Story of a Tsunami

Lafcadio Hearn was born in Greece in 1850. At the age of two he was brought to Ireland, and at nineteen he immigrated to the United States, where he settled in Cincinnati, Ohio, to become a journalist. At the age of forty he travelled to Japan, where he gained a teaching post at a Middle School and then became a professor at Waseda University. The reports he wrote familiarized the world with life in pre-industrial Meiji era Japan, including the island country's experiences with the sea and its sometimes very destructive storms.

Here is an extended quote from his story, *A Living God*:

> From immemorial time the shores of Japan have been swept, at irregular intervals of centuries, by enormous tidal waves, — tidal waves caused by earthquakes or by submarine volcanic action. These awful sudden risings of the sea are called by the Japanese tsunami. The last one occurred on the evening of June 17, 1896, when a wave nearly two hundred miles long struck the north-eastern provinces of Miyagi, Iwate, and Aomori, wrecking scores of towns and villages, ruining whole districts, and destroying nearly thirty thousand human lives. The story of Hamaguchi Gohei is the story of a like calamity which happened long before the era of Meiji, on another part of the Japanese coast.

He was an old man at the time of the occurrence that made him famous. He was the most influential resident of the village to which he belonged: he had been for many years its muraosa, or headman; and he was not less liked than respected. The people usually called him Ojiisan, which means Grandfather; but, being the richest member of the community, he was sometimes officially referred to as the Choja. He used to advise the smaller farmers about their interests, to arbitrate their disputes, to advance them money at need, and to dispose of their rice for them on the best terms possible.

Hamaguchi's big thatched farmhouse stood at the verge of a small plateau overlooking a bay. The plateau, mostly devoted to rice culture, was hemmed in on three sides by thickly wooded summits. From its outer verge the land sloped down in a huge green concavity, as if scooped out, to the edge of the water; and the whole of this slope, some three quarters of a mile long, was so terraced as to look, when viewed from the open sea, like an enormous flight of green steps, divided in the centre by a narrow white zigzag, a streak of mountain road. Ninety thatched dwellings and a Shinto temple, composing the village proper, stood along the curve of the bay; and other houses climbed straggling up the slope for some distance on either side of the narrow road leading to the Choja's home.

One autumn evening Hamaguchi Gohei was looking down from the balcony of his house at some preparations for a merry-making in the village below. There had been a very fine rice-crop, and the peasants were going to celebrate their harvest by a dance in the court of the ujigami. The old man could see the festival banners (nobori) fluttering above the roofs of the solitary street, the strings of paper lanterns festooned between bamboo poles, the decorations of the shrine, and the brightly colored gathering of the young people. He had nobody with him that evening but his little grandson, a lad of ten; the rest of the household having gone early to the village. He would have accompanied them had he not been feeling less strong than usual.

The day had been oppressive; and in spite of a rising breeze, there was still in the air that sort of heavy heat which, according to the experience of the Japanese peasant, at certain seasons precedes an earthquake. And presently an earthquake came. It was not strong enough to frighten anybody; but Hamaguchi, who had felt hundreds of shocks in his time, thought it was queer, — a long, slow, spongy motion. Probably it was but the after-tremor of some immense seismic action very far away. The house crackled and rocked gently several times; then all became still again.

As the quaking ceased Hamaguchi's keen old eyes were anxiously turned toward the village. It often happens that the attention of a person gazing fixedly at a particular spot or object is suddenly diverted by the sense of something not knowingly seen at all, — by a mere vague feeling of the unfamiliar in that dim outer circle of unconscious perception which lies beyond the field of clear vision. Thus it chanced that Hamaguchi became aware of something unusual in the offing. He rose to his feet, and looked at the sea. It had darkened quite suddenly, and it was acting strangely. It seemed to be moving against the wind. It was running away from the land. Within a very little time the whole village had noticed the phenomenon. Apparently no one had felt the previous motion of the ground, but all were evidently astounded by the movement of the water. They were running to the beach, and even beyond the beach, to watch it. No such ebb had been witnessed on that coast within the memory of living man. Things never seen before were unfamiliar spaces of ribbed sand and reaches of weed-hung rock were left bare even as Hamaguchi gazed. And one of the people below appeared to guess what that monstrous ebb signified.

Hamaguchi Gohei himself had never seen such a thing before; but he remembered things told him in his childhood by his father's father, and he knew all the traditions of the coast. He understood what the sea was going to do. Perhaps he thought of the time needed to send a message to the village, or to get the priests of the Buddhist temple on the hill to sound their big bell .... But it would take very much longer to tell what he might have thought than it took him to think. He simply called to his grandson — "Tada ! — quick, — very quick! ... Light me a torch.." Taimatsu, or pine-torches, are kept in many coast dwellings for use on stormy nights, and also for use at certain Shinto festivals. The child kindled a torch at once; and the old man hurried with it to the fields, where hundreds of rice-stacks, representing most of his invested capital, stood awaiting transportation. Approaching those nearest the verge of the slope, he began to apply the torch to them, — hurrying from one to another as quickly as his aged limbs could carry him. The sun-dried stalks caught like tinder; the strengthening sea-breeze blew the blaze landward and presently, rank behind rank, the stacks burst into flame, sending skyward columns of smoke that met and mingled into one enormous cloudy whirl. Tada, astonished and terrified, ran after his grandfather, crying, "Ojiisan! why? Ojiisan! — why?"

But Hamaguchi did not answer: he had no time to explain; he was thinking only of the four hundred lives in peril. For a while the child stared wildly at the blazing rice, then burst into tears and ran back to the house, feeling sure that his grandfather had gone mad. Hamaguchi went on firing stack after stack, till he had reached the limit of his field; then he threw down his torch and waited. The acolyte of the hill-temple, observing the blaze, set the big bell booming; and the people responded to the double appeal. Hamaguchi watched them hurrying in from the sands and over the beach and up from the village, like a swarming of ants and, to his anxious eyes, scarcely faster; for the moments seemed terribly long to him. The sun was going down; the wrinkled bed of the bay, and a vast sallow speckled expanse beyond it, lay naked to the last orange glow; and still the sea was fleeing toward the horizon.

Really, however, Hamaguchi did not have very long to wait before the first party of succor arrived, — a score of agile young peasants, who wanted to attack the fire at once. But the Choja, holding out both arms, stopped them. "Let it burn, lads!" he commanded, "let it be! I want the whole mura here. There is a great danger, — taihen da!"

The whole village was coming; and Hamaguchi counted. All the young men and boys were soon on the spot, and not a few of the more active women and girls; then came most of the older folk, and mothers with babies at their backs, and even children, — for children could help to pass water; and the elders too feeble to keep up with the first rush could be seen well on their way up the steep ascent. The growing multitude, still knowing nothing, looked alternately, in sorrowful wonder, at the flaming fields and at the impassive face of their Choja. And the sun went down.

"Grandfather is mad, — I am afraid of him!" sobbed Tada, in answer to a number of questions. "He is mad. He set fire to the rice on purpose: I saw him do it!" "As for the rice," cried Hamaguchi, "the child tells the truth. I set fire to the rice . . . Are all the people here?"

The Kumi-cho and the heads of families looked about them, and down the hill, and made reply: "All are here, or very soon will be . . . We cannot understand this thing." "Kita!" shouted the old man at the top of his voice, pointing to the open. "Say now if I be mad!"

Through the twilight eastward all looked, and saw at the edge of the dusky horizon a long, lean, dim line like the shadowing of a coast where no coast ever was, — a line that thickened as they gazed, that broadened as a coast-line broadens to the eyes of one approaching it, yet incomparably more quickly. For that long darkness was the returning sea, towering like a cliff, and coursing more swiftly than the kite flies.

"Tsunami!" shrieked the people; and then all shrieks and all sounds and all power to hear sounds were annihilated by a nameless shock heavier than any thunder, as the colossal swell smote the shore with a weight that sent a shudder through all the hills, and a foam-burst like a blaze of sheet-lightning. Then for an instant nothing was visible but a storm of spray rushing up the slope like a cloud; and the people scattered back in panic from the mere menace of it. When they looked again, they saw a white horror of sea raving over the place of their homes. It drew back roaring, and tearing out the bowels of the land as it went. Twice, thrice, five times the sea struck and ebbed, but each time with lesser surges; then it returned to its ancient bed and stayed, — still raging, as after a typhoon.

On the plateau for a time there was no word spoken. All stared speechlessly at the desolation beneath, — the ghastliness of hurled rock and naked riven cliff, the bewilderment of scooped-up deep-sea wrack and shingle shot over the empty site of dwelling and temple. The village was not; the greater part of the fields were not; even the terraces had ceased to exist; and of all the homes that had been about the bay there remained nothing recognizable except two straw roofs tossing, madly in the offing. The after-terror of the death escaped and the stupefaction of the general loss kept all lips dumb, until the voice of Harnaguchi was heard again, observing gently, — "That was why I set fire to the rice."

He, their Choja, now stood among them almost as poor as the poorest; for his wealth was gone, but he had saved four hundred lives by the sacrifice. Little Tada ran to him, and caught his hand, and asked forgiveness for having said naughty things. Whereupon the people woke up to the knowledge of why they were alive, and began to wonder

> at the simple, unselfish foresight that had saved them; and the headmen prostrated themselves in the dust before Hamaguchi Gohei, and the people after them.
>
> Then the old man wept a little, partly because he was happy, and partly because he was aged and weak and had been sorely tried. "My house remains," he said, as soon as he could find words, automatically caressing Tada's brown cheeks; "and there is room for many. Also the temple on the hill stands; and there is shelter there for the others." Then he led the way to his house; and the people cried and shouted.

The recorded history of 18 notable tsunami in modern times goes back to the eighteenth century, beginning with 1755 in Lisbon, Portugal, which caused 30,000 deaths. The tsunami with the highest wave recorded in the nineteenth century, of 35 m,[1] seems to have been the one in 1883 at the site of the Krakatau volcano in Indonesia, and caused 36,000 deaths. The one in 2004 that hit Indonesia, Sri Lanka and India caused a total of about 245,000 deaths. Of course the wave of most recent memory was the tsunami that struck Tohoku, Japan, in 2011, reaching a height of about 40.5 m, and causing almost 16,000 deaths and severe damage to a nuclear power plant, which contaminated a large area with radioactivity. There were many more tsunami of lesser size which we shall not enumerate.

The enormous, destructive waves of tsunami fortunately occur only rarely, but we want to try to understand them. By this we mean not only describing their causes but, if possible, finding their underlying equations, because mathematical equations serve the important purpose of explaining observations. What do we mean by such a statement? Explaining observed phenomena by equations implies that their precise description follows logically from the equations' solutions. There is no need for further postulates that have to be accepted on faith; only the use of straightforward logic. If the

[1] 1 meter, abbreviated m, equals 3.28 feet.

accurate description of a given set of phenomena is a solution of a given equation, or set of equations — together with the mathematical conditions called boundary conditions or initial conditions, if such are needed to make the solution unique — then we regard these equations as the explanation of the phenomena. Let us therefore turn to the history of this explanatory method, followed by its application to other more familiar kinds of waves. We shall then return to tsunami to examine whether there is an equation, or a set of equations, that serves the same purpose for them.

2

Science and Mathematics

Physics is notorious for using a lot of mathematics, and many people are scared of it. Why is that so, and how did it come about that mathematical equations are used by physicists to explain phenomena? Let's look at some of the history of both science and mathematics to understand this better.

Living in the sixth century BCE, Thales of Miletos was the first influential thinker who could be called a scientist in the modern sense of that word. For any natural puzzle presented to him, he was always searching for an explanation based on fundamentals, never satisfied with superficial tricks. Both the first mathematician and the first astronomer in Greece, he was born in the wealthy Ionian harbor city of Miletos and was always included in the group of seven legendary wise men of the early Greek tradition, mostly on the basis of an apocryphal story that he had correctly predicted the solar eclipse in the spring of 585, the darkness of which so scared the leaders of the battling Lydian and Persian armies of the wrath of the gods that they instantly made peace. Superior knowledge was for once used for peaceful ends. Benefiting from his extensive travels in the much more civilized and advanced country of Egypt, Thales is also credited with enabling King Croesus's army to cross the river Halys safely by diverting it, thereby demonstrating that the gods did not govern the flow of rivers.

Often using geometry to solve such problems as to estimate accurately the distance of ships visible from shore, or the height of buildings, Thales then always searched for general principles underlying his solutions, even though we do not know to what extent he was actually able to prove these principles and theorems. The man had an enormous benefitial influence on ancient Greek culture as well as on our own.

Pythagoras, about whose life relatively little is reliably known, except that he was born on the island of Samos near Miletos, may have been a student of Thales, who is said to have recognized his genius quite early. After much travel in Babylon and Egypt he eventually settled at Croton, near the southern tip of Italy, then a Dorian colony. There he established an influential cult of religious mystics, surrounding himself with disciples who shared his secrets and dietary

Figure 2.1. Pythagoras depicted in the center by Raphael in the school of Athens, of which this is a detail.

taboos, all living simply and poorly. His cult continued to exist for about fifty years after his death in about 500 BCE, but was then suppressed because it became too political.

In the tradition of Thales, Pythagoras too was always searching for general theorems underlying specific mathematical solutions to problems. A prime example, of course, is his well-known theorem still taught in high schools today: the square of the hypotenuse of a right triangle equals the sum of the squares of the other two sides, which he proved by simple geometry. Without knowledge of the general theorem, special cases had been used for a long time for purposes of laying out rectangular agricultural fields: for example, in a triangle formed by strings of lengths 3, 4, and 5 units, the angle opposite the longest string is 90°, a very useful fact to know if you want to draw a rectangle.

Since you may not have been taught the proof of the theorem, here it is. For a given right triangle EGH, construct the square ABCD whose side-length equals the sum of the lengths EG + GH, as in Figure 2.2. Then, if (ABCD) denotes the area of the figure enclosed

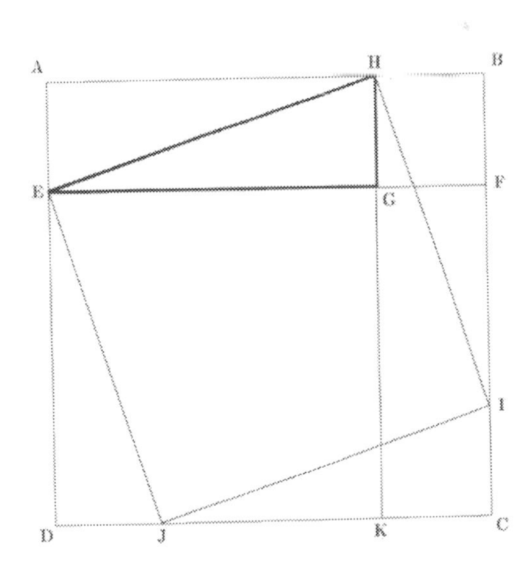

Figure 2.2. Proof of the theorem of Pythagoras.

by ABCD,

$$(ABCD) = (EGKD) + (HBFG) + 2(AHGE) = (EHIJ) + 4(AHE).$$

But $(AHGE) = 2(AHE)$; therefore $2(AHGE)$ cancels $4(AHE)$, and what remains is $(EHIJ) = (EGKD) + (HBFG)$. q.e.d.

The entire way of thinking of Pythagoras was based on numbers and mathematical equations, which he regarded as the essence of the world. We surely have inherited some of his philosophy.

The two best known ancient Greek philosophers, of course, are Plato and Aristotle. Although the gate to his Academia was inscribed "Let no one enter here who is ignorant of mathematics," Plato's influence on science was not fruitful. That's because his attention was never on the world as it is actually observed but as he thought it ought to be ideally. The intellectual orientation of his disciple, Aristotle, who lived in the fourth century BCE, and of the Lyceum he founded, were entirely different, directed toward fact-collecting and experimentation.

For our purposes, what is of interest in Aristotle's philosophy is his law governing motion of ordinary objects on the surface of the Earth, such as oxen hauling heavy burdens. To move an object of mass M a distance D, he said, requires the application of a force F for a time T such that $F \times T$ is proportional to $M \times D$. Since D/T, the distance traveled divided by the time it took, is the average velocity of the moving object, his law of motion was later interpreted as decreeing that the required force F was proportional to the product $M \times V$, or $F = R \times M \times V$, where R is the resistance. This being such an eminently reasonable proposition that agrees with our experience in a world in which friction and other resistance to motion are ubiquitous, it was easily accepted and is still regarded as obvious by most people. The law implied, of course, that if it were possible to produce a vacuum, in which there would be motion with no friction, any force would move an object with infinite speed. From this he concluded that here on Earth, a vacuum had to be impossible. (Since Aristotle postulated completely different laws of motion for objects as far away as the Moon and beyond, where he thought the laws of physics did not apply, there was no reason why there could not be a vacuum in the translunar region.)

Aristotle's equation $F = R \times M \times V$, remained the accepted law of motion for more than a millennium and a half until it was challenged by the English 'doctor profundus' Thomas Bradwardine. Educated at Balliol College, Oxford, with a degree in theology but at the same time also a skilful mathematician, Bradwardine was born about 1290, became professor of divinity and chancellor of Oxford university and consecrated in 1349 as Archbishop of Canterbury, head of the Church of England. Forty days after his consecration, however, he died of the plague and was buried in Canterbury.

Bradwardine accepted Aristotle's view that the velocity with which an object moved was deteremined by the force acting on it and the resistance it experienced. However, he believed that when the force is gradually diminished or the resistance increased until the two became equal, all motion should stop, which Aristotle's equation did not predict. He therefore changed the law so that a doubling of the velocity required not a doubling of the ratio of force to resistance but rather this ratio squared. In modern mathematical language, he postulated that the equation of motion be of the form $F/R = (\text{const.})^V$ rather than $F/R = \text{const.} \times V$, as Aristotle had it. His knowledge of math was sophisticated enough to understand that this form of the equation of motion implied that if $F/R = 1$, $V = 0$ would follow. His new equation still implied that if $R = 0$, as in a vacuum, an infinite velocity would follow; so there could be no vacuum on the surface of the Earth, just as Aristotle had concluded.

Aristotle's equation of motion, either in its original or in its altered form, remained as the accepted law for another three hundred years until the scientific revolution of the seventeenth century. After Galileo's experiments, it was Isaac Newton who introduced an entirely different perspective. Whereas Aristotle's primary attention was on motion on Earth, where there was always resistance, Newton's was on outer space, where the planets moved without resistance and the existence of a vacuum was assumed as a fact. There should be only one fundamental law of motion, the same for the planets and for balls thrown on Earth. Resistance was for him a secondary complication to be added, when needed — as a 'perturbation' in more modern language. Newton's universal equation of motion, replacing Aristotle's and Bradwardine's, as expressed later by the great Swiss mathematician Leonard Euler, was $F = M \times A$, where A is the acceleration of the object of mass M. (Newton never wrote his law of motion in this now-so-familiar form; he tended to think geometrically rather than algebraically.) This was a fundamental change — as every teacher of elementary physics knows, it is still hard to accept for many students, natural-born Aristotelians: the applied force does not determine the velocity with which an object moves but its acceleration, i.e. the change in its velocity. To move with a constant velocity requires no force.

The equations expressing Newton's laws of motion in their most general form were next applied to any number of objects exerting forces on one another.[1] This included his universal force of gravitation between any two objects, proportional to the product of their masses, acting at a distance and decreasing as the square of that distance. All this complicated mathematics turned out to explain not only the fact, demonstrated by Galileo, that all objects on Earth fall with the same acceleration irrespective of their mass (since the downward force is proportional to the object's mass, its mass cancels out in its equation of motion), but also the form of the orbits of all the planets, including that of the Earth, in all of their known details.

The Polish astronomer Nicolaus Copernicus had roiled Christian Europe and angered the Vatican — Galileo had been tried and

[1] These equations are differential equations: he had invented the differential calculus specifically for this purpose. If you don't understand what a differential equation is it doesn't really matter, but you may find a simple introduction in my book *What Makes Nature Tick?*

condemned by the Inquisition for advocating the Copernican view, a verdict that was not rescinded until 1992 — by replacing the old Ptolemaic geocentric system with a solar-centric system, postulating that all the planets, including the Earth, orbited the Sun in circles.

Born in 1473 in the German-speaking town Torun (Thorn) south of Gdansk, in the Kingdom of Poland, a son of a merchant, Copernicus was first educated at the University of Kraków, learning primarily mathematics and astronomy as well as reading much of Aristotle's philosophy. He then attended the Universities of Bologna and Padua, being introduced to the writings of Pico della Mirandola and Regiomontanus's critical translation *Epitome of Ptolemy's Almagest* and learning from them that astronomers had always disagreed among themselves about the order of the planets and that there were glaring discrepancies between Ptolemy's geocentric model of the solar system and actual observations. After receiving a doctorate in canon law at the University of Ferrara, he obtained a sinecure at Wroklaw as well as a canonship at Frombork, where he built himself his own little astronomical obervatory, though he rarely actually used it. He died in Frombork is 1543.

Copernicus first challenged the astronomical views of Aristotle and Ptolemy in his *Commentariolus*, in which he sketched in preliminary form, without any claim of priority, a heliocentric planetary system with the Earth spinning on its own axis as well as orbiting the Sun in a circle, and all the planets circling the Sun in circles as well. This work was published only in fragmentary form. All the details were later given in *De revolutionibus orbium velestium libri vi*, a work he was extremely reluctant to have printed for 36 years, until the year before his death. (The earliest proponent of a heliocentric planetary system, Aristarchus of the 3rd century BCE, had been accused of impiety, as Copernicus well knew.)

Even though many of its predictions did not really agree very well with astronomical observations, this new model of the solar system, much simpler than that of Ptolemy, was readily accepted in Asian countries like China and India because it led to a more accurate calendar, but it was considered revolutionary in Christian Europe as it removed the Earth with its biblical history from the center of the universe.

A century later the German astronomer Johannes Kepler had modified the Copernican system. Born in 1571 in Weil der Stadt near Stuttgart in Germany, Kepler studied philosophy and theology at the University of Tübingen, though he was greatly interested in astronomy and excelled in mathematics. His first position was as a teacher of those subjects at a school that later became the University of Graz, where he publicly defended the Copernican system while trying to construct Platonic models of the planetary orbits by means of three-dimensional polyhedra, an endeavor that led nowhere. After meeting the great astronomer Tycho Brahe, who was in the process of building a new observatory near Prague, Kepler, a religious Lutheran, had to leave Graz because he refused to convert to Catholicism, and he moved to Prague. When Tycho unexpecctedly died the following year, 1601, Kepler was appointed his successor as the imperial mathematician at the court of Emperor Rudolph II. He died in 1630 in Regensburg, Bavaria.

Beginning under Tycho's direction, Kepler paid particular attention to the motion of the planet Mars, coming to the conclusion that its orbit could not be circular but had to be ovoid in shape. More precise observations showed that the 'ovoid' orbit was an ellipse, with the Sun at one of its foci. What is more, he found that Mars did not move with a uniform speed but, when closer to the Sun, it moved faster than when farther away. As he was a very precise observer as well as mathematically inclined, he formulated what he discovered by saying that the straight line from the Sun to the planet traces out equal areas in equal times. He found these same observations also to hold for the other planets, and they became known as the first two of Kepler's three laws of planetary motion. The third law he formulated was that the squares of the periods of the planets — the squares of the lengths of their years — are proportional to the cubes of their

Figure 2.3. The back of a 10-euro coin minted in 2002.

mean distance from the Sun. Copernicus had been the revolutionary, but Kepler got it right.

Though hailing the image of the solar system he had unveiled — many others, including Galileo, however, regarded his ellipses as ugly and much preferred the Copernican circles — he could of course not explain exactly why the planets moved the way they did. That explanation was not provided until after Kepler's death by Isaac Newton, the man who personified the scientific revolution that transformed the world in the seventeenth ceentury. Triumphantly, Newton was able to explain Kepler's observational facts by showing that they followed as consequences of the equations expressing his laws of motion as well as his law of universal gravitation. Thereafter there could be no doubt that Newton's laws were correct.

According to the Julian Callendar still in use in England at the time, Isaac Newton was born prematurely on Christmas Day of 1642, the year of Galileo's death, at Woolsthorpe in Lincolnshire. His father, a prosperous farmer, had died three months earlier, and when he was three his mother remarried, leaving Isaac in the care of his maternal grandmother, in a house lacking all affection. His neurotic, tortured personality manifesting itself later all his life can clearly be traced to his upbringing. (He may also have been afflicted with Asperger syndrome.) His early education, from twelve to seventeen, was at the King's School, Grantham, where the curriculum consisted almost entirely of Bible studies and Latin, after which his mother, widowed again, attempted to make him a farmer — he hated 'the idiocy of rural life'[2] — but was persuaded to let him return to school. At the age of eighteen he was admitted to Trinity College, Cambridge, as a sub-sizar, which meant that he had to perform chores for the more affluent students and for fellows.

Cambridge University at the time was still dominated by a stultifying Aristotelianism, teaching not merely his sound views but everything the philosopher had said, even when based on incorrect data. Newton, rejecting Aristotle, studied instead the works of Galileo, Descartes, and the French philosopher and mathematician Pierre Gassendi. Filling his notebooks with searching questions, he was always looking for the experimental or observational consequences of any theory, some of which he answered by experimenting himself, even on his own body, looking at the sun or slipping a bodkin 'betwixt my eye & ye bone as near to ye backside of my eye as I could' to test his ideas about light, vision, and color. As the university at that time badly neglected mathematics until they established the Lucasian Chair in Mathematics in 1663, he became almost an autodidact in the subject which began to interest him intensely. His study of Euclid, the ancient Greek, would greatly influence his own style and methodology. Graduating with a Batchelor of Arts degree in 1665 he returned to Woolsthorpe for eighteen months as the university closed down because Cambridge, along with London and surroundings, were hit by the bubonic plague, the famous Black Death.

The years 1664 to 1666 are sometimes referred to as the *anni mirabiles* of Newton's life, as that was the period of his most fecund

---

[2]This somewhat mistranslated phrase was coined some two centuries later by Marx and Engels in their *Communist Manifesto.*

scientific work. Sleeping and eating erratically during those two solitary years, he laid the foundations of his invention of the calculus (he called it the method of fluxions), of his laws of motion, the universal law of gravitation, and of his discoveries in optics. The law of gravitation he discovered, he said when asked later, 'by thinking on it continually.'

In 1666 he completed three mathematical papers on differential and integral calculus, including what is now known as 'the fundamental theorem of the calculus,' which states that the differential and integral calculus are one another's inverses. (These papers, however, remained unpublished.) A powerful generalization of hints contained in Archimedes's 'method of exhaustion' and in the infinitesimals introduced by the French mathematician Pierre de Fermat, the calculus would eventually blossom into a very large area of mathematics now called 'analysis.' All his work in mathematics was directly motivated by his detailed analysis of physical problems of motion.

Returning to Cambridge, Newton became a Fellow of Trinity College and received the Master of Arts degree in 1668. At that time, Isaac Barrow, the first incumbent of the Lucasian Chair in Mathematics, was the only one to whom Newton had shown the three unpublished path-breaking papers he had written, and when he read them he promptly vacated the Chair to enable Newton to occupy it at the age of twenty-six.

His greatest work, the monumental *Philosophae Naturalis Principia Mathematica*, written entirely in Latin, he presented to the Royal Society in 1686, but because of shortage of funds it was not published until the following year when his friend, the astronomer Edmund Halley paid for its printing. After that he became embroiled in a number of unpleasant and protracted priority disputes that showed the prickly side of his character and his extreme sensitivity to any hint of criticism. One of these feuds involved Robert Hooke, a distinguished scientist and co-founder of the Royal Society, who claimed to have thought of explaining the planetary orbits by means of an inverse-square law and accused Newton of plagiarism. (Newton had indeed given no credit to Hooke's somewhat confused work, even though he was well aware of it.)

The most famous of Newton's priority quarrels, however, was that with the German philosopher and mathematician Gottfried Wilhelm Leibniz, who had independently invented the calculus ten years after Newton's paper of 1665 that had remained unpublished and seen only by Barrow. Newton, while remaining neurotically secretive about his own work, publicly accused Leibniz of plagiarism. The after-effect of this unfair treatment of the well-known philosopher was a long-lasting alienation between British and continental European mathematicians.

After a period of severe depression, Newton lost interest in physics and mathematics, was appointed warden of the Mint and knighted by Queen Anne. Sir Isaac Newton died in 1727 and was accorded a state funeral at Westminster Abbey. While about two centuries later the applicability of his laws of motion, now called classical mechanics, as well as of his law of universal gravitation — all his equations — were found to be limited to objects that are not submicroscopic nor too massive and whose relative motion does not approach the speed of light, they can be expected to remain valid and useful within that range of applicability forever.

The next set of physical phenomena that became successfully 'mathematicised' was that of electricity and magnetism. Electrostatic attractions had been know since antiquity. Thales of Miletos is generally credited with the discovery that, after being rubbed, a piece of amber will attract small objects. In the early eighteenth century, the French infantry officer Cisternay Dufay showed that all solid objects other than metals could be electrified by rubbing, some attracting one another, others repelling, and that metals were able to communicate such 'electrical virtues' differently from other substances. As the English amateur astronomer Stephen Gray had recently found, such 'electrical virtues' could also sometimes be communicated through the air, occasionally with a spark. In the eighteenth century, when electricity was a fashionable topic of intellectual discourse, Benjamin Franklin famously demonstrated that lightening is an electric phenomenon by dangling an iron key from a kite in a thunderstorm (which led to his invention of the lightning rod).

That a lodestone attracts iron was mentioned by Lucretius in the first century BCE, and both the Chinese and Western Europeans used the magnetic compass for navigation in the twelfth century. As time went on, investigations of these strange forces, sometimes attractive and sometimes repulsive, continued over the centuries. It was not until the nineteenth century, however, that the apparently quite different electric and magnetic forces were found to be somehow related. The Danish physicist Hans Christian Oersted discovered in 1820 that a wire carrying an electric current could deflect a compass needle: it evidently produced magnetism. After the French physicist and chemist André Marie Ampère witnessed a demonstration of Oersted's discovery, he experimented further and found that two parallel current-carrying wires attracted or repelled one another, depending on whether the currents in them ran in the same or opposite direction.

Born in the parish of St. Nizier near Lyon in 1775, his father a successful businessman and admirer of the philosophy of Jean-Jacques Rousseau, André Marie Ampère was mostly schooled at home, becoming a polymath. After his father was guilliotined during the French Revolution, he became a mathematics teacher and subsequently was appointed professor of physics and chemistry at the École Centrale at Bourg-en-Bresse and eventually moved to Paris where he was elected to a chair in experimental physics at the Collège de France. He died in 1836 in Marseille. Today we measure electric currents in units of ampere, and the pysical law that relates the magnetism produced by an electric current at a given distance to the strength of that current is called Ampère's law.

There is however also the inverse effect, called induction: magnetism can produce electric currents. This was discovered by the

American physicist Joseph Henry and, independently, by the English chemist and physicist — though he always called himself a natural philosopher — Michael Faraday.

Henry was born in 1797 in Albany, New York, the son of a laborer who had immigrated from Scottland. As his father died when he was young, he was brought up by his grandmother and apprenticed to a watchmaker, but his interest in science was sparked by independent reading. After attending the Albany Academy, having been given free tuition, he was appointed assistant engineer for the survey for the construction of a State road to Lake Erie. At the age of 28 he was appointed professor of mathematics and natural philosphy at the Albany Academy. His curiosity centered mostly on terrestrial magnetism, he began experimenting with magnetism in general and constructed an improved electromagnet — a primitive form of such an object had first been built by the English inventor William Sturgeon — by tightly winding a insulated wire coil around a core made of soft iron which does not retain magnetism. So long as there was an electric current running through the wire the iron core was magnetic: an electromagnet that could be turned on and off. He also made a number of practical inventions, worked in astronomy and aeronautics and served as the first Secretary of the Smithsonian Institution. Joseph Henry died in 1878 in Washington, D.C.

The son of a blacksmith, Michael Faraday was born in 1791 in Newington, Surrey, England, and self-educated by reading the *Encyclopedia Britannica*; he was sent to London by his father as an apprentice to a book binder. Young Faraday had the good fortune to have his extraordinary talent discovered by Humphry Davy, who was considered the greatest chemist in Europe at the time and for whom Faraday began to work as a temporary assistant. Davy took Faraday with him on an extended tour of the continent, where he met and talked to many of the most prominent scientists of the day, a voyage that served the young man as the substitute for a university education. On his return he obtained a position at the Royal Institution, where he remained to become the greatest experimenter in physics and chemistry of the nineteenth century, as well as an extremely popular lecturer to public audiences. When he retired, Queen Victoria provided him with a furnished apartment at Hampton Court, where he died in 1867.

Among Faraday's many disoveries was the fact that electric currents produce magnetism, a fact now called Faraday's law of induction. Combined with Ampère's law, induction showed that electricity and magnetism were really one set of phenomena: electromagnetism. Faraday now tried to understand how this electromagnetic force could be transmitted through space without using the intellectually repulsive action at a distance Newton had postulated for the force of gravity. Imaginative as he was, he came up with one of the most fruitful and far-reaching ideas in the history of physics: the electromagnetic field. An electric charge at one point in space did not directly act on another charge at a distance, nor did magnets, but they produced a condition of space — the field — that then acted on space at neighboring points, influencing, extending, and travelling everywhere. He envisioned all of space filled with this force-field like rubber bands. However, his mathematical power was insufficient to formulate this extremely fruitful concept in a rigorous manner. Such a formulation had to wait some forty years for the Scottish physicist James Clerk Maxwell.

As a child, Maxwell was said to have had an unquenchable curiosity. He had been born in 1831 in Edinburgh, his father a well-to-do lawyer, brother of the 6th Baronet of Clerk of Penicuik; after his mother died when he was eight years old, and the hired tutor, who regarded him as a slow learner, was dismissed, his father sent him to the Edinburgh Academy, where the young boy was regarded by the others as a country bumpkin. As a student he developed a passion for drawing and was especially facsinated by geometry, winning the school's mathematical medal at the age of 13. At the age of 14 he published his first scientific paper, which was presented to the Royal Society of Edinburgh by a professor because the author was deemed too young.

Two years later Maxwell entered the University of Edinburgh, where he remained for three years, moving on to Cambridge University in 1850, initially to Peterhouse but soon transferring to Trinity College, where he was elected to the elite secret society called the Cambridge Apostles. Graduating with a degree in mathematics in 1854, he was made a Fellow at Trinity the next year. Appointed to a Chair of Natural Philosophy at Marischal College, Aberdeen, he left Cambridge in 1856. In 1860, when Marischal College merged with King's College to form the new University of Aberdeen, he was appointed to the professorship of natural philosophy at King's College, London, and a year later he was elected to the Royal Society.

Maxwell showed his interest in Faraday's ideas on electromagetism as early as 1856, when he read his paper 'On Faraday's lines of force' to the Cambridge Philosophical Society. But the full range of his mathematical formulation of a complete theory of the electromagnetic field had to wait for the completion of his textbook *A Treatise on Electricity and Magnetism*, published in 1873. This work was followed by many other important contributions to physics. James Clerk Maxwell died at the age of 48 in 1879 in Cambridge. He and Charles Darwin were undoubtedly the most influential scientists of the nineteenth century.

Being an inveterate model builder, he constructed the entire edifice of his theory of elecromagnetism on the basis of an elborate mechanical model of the ether. This led him to a set of laws formulated as differential equations that had to be satisfied by the electric and magnetic fields, now always simply refered to as *Maxwell's equations.*[3] These equations contained only one constant, a number inherited from Ampère's and Faraday's laws, the ratio of the different units employed by the measurements of static electric charges and of electric currents. (Significantly, the two experimenters

---

[3]These are laws we now call 'classical.' The quantum revolution of the twentieth century limited their area of applicability just as it did Newton's laws of motion. This need not concern us here.

William Weber and Rudolph Kohlrausch had discovered that this ratio was equal to the velocity of light.)

When Heinrich Hertz, an experimenter who verified many of the predictions implied by Maxwell's work, later wrote his own book *Electric Waves*, he very simply wrote 'Maxwell's theory is Maxwell's system of equations.' The underlying model Maxwell had used was simply discarded and forgotten; like Lewis Carroll's Cheshire cat, it disappeared, leaving only its grin behind, the equations.

Again we have here an important example of the use of a set of equations (with their needed boundary conditions for specific instances) as explanations of the enormous variety of observed electric and magnetic phenomena. Electromagnetic waves will be among the phenomena we shall now turn to in order to compare them to the properties of tsunami. First, however, we shall consideer the gentle waves of music. Though their effects are quite familiar to us, they were not recognized as waves for many centuries even as many of their qualities were studied very successfully.

3

The Sound of Music

We have already discussed in Chapter 2 the Greek philosopher Pythagoras, whose view of the world was fundamentally based on mathematics and that numbers and mathematical equations described the essence of the world. And this notion he applied even to his ideas about music, which interested him greatly. Legend has it that, listening to the hammering of a blacksmith on his anvil he noted the harmony of the sounds and decided that they were most harmonious when the weights of the hammers were in certain simple ratios like 1/2 or 2/3, etc. While this story is certainly apocryphal, as there is no relation between musical harmony and the weight of hammers causing the sounds, Pythagoras probably did discover that harmony of sound emitted by string instruments is indeed related to the simple ratios of string lengths.

It took well over two thousand years and the scientific revolution for the relation between the pitch of musical notes emitted by string instruments and the frequency of the vibrations of their strings to be understood. The first person to have done so seems to have been a younger contemporary of Galileo, the French natural philosopher, mathematician, and theologian Marin Mersenne.

Though Mersenne was obsessed with scientific precision, the first to actually apply Newton's new equations to the motion of a vibrating string was the English mathematician Brook Taylor at St. John's College in Cambridge in 1715, followed about forty years later by the Dutch–Swiss mathematical physicist Daniel Bernoulli at the University of Basel. The result was that the shape of a gently plucked string of mass M per unit length, and tension T is governed by what is now called the wave equation. (Technically, the wave equation is a simple linear, homogenious partial differential equation, but if you don't know what that means, it doesn't matter.) It contains but one parameter c_{st} (which is related to M and T by $c_{st}^2 = T/M$). This equation has infintely many solutions of the form[1]

$$A \sin (2\pi x/\lambda) \cos (2\pi tf),$$

where the constants f and λ are related by $f\lambda = c_{st}$. As the functions sin and cos are periodic with period 2π (see Fig. 3.1), λ and f denote, respectively, the wavelength and the frequency of vibration of such a solution of the wave equation. The constant A is arbitrary; its value is determined by the strength with which the string is plucked or hit — as in a piano.

The ends of the string in a musical instrument being fixed, the wavelength λ has to be an integral fraction of twice the length L of the string, $\lambda = 2L/n$, $n = 1, 2, 3, \ldots$, which makes the solution equal to zero at both ends. (The sine function vanishes both at π and at $\pi/2$, see Fig. 3.1.) The first two possibilities of such a 'standing wave' are shown in Fig. 3.2. The vibration frequencies of these solutions are consequently $f = c_{st}/\lambda = n(c_{st}/2L)$, $n = 1, 2, 3, \ldots$. The lowest frequency of a string of length L is thus $f = c_{st}/2L$, but it also produces vibrations with a frequency twice as high as the lowest, called the first harmonic, another with a frequency three times as high as the lowest, i.e., whose frequency ratio to the first harmonic is 3:2, called a *fifth* by musicians, etc. (These vibrations are also known

---

[1]Remember from your trigonometry class: sin (α) is the ratio, in a right triangle, of the length of the side opposite to α to that of the hypothenuse, and cos (α) is the ratio of the length of the side next to α to that of the hypothenuse. Angles are expressed in radians, so that 180° corresponds to π; α and λ are the Greek letters alpha and lambda.

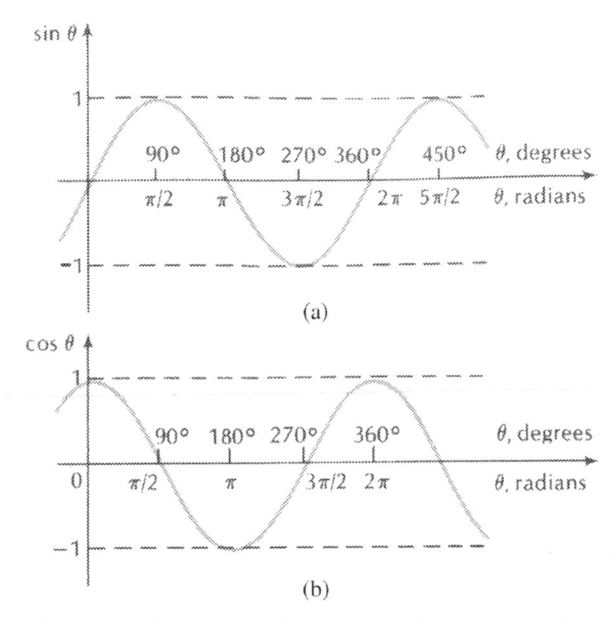

Figure 3.1. This shows the shapes of sine and cosine waves. Both extend with the same period to infinity on the right and minus infinity on the left.

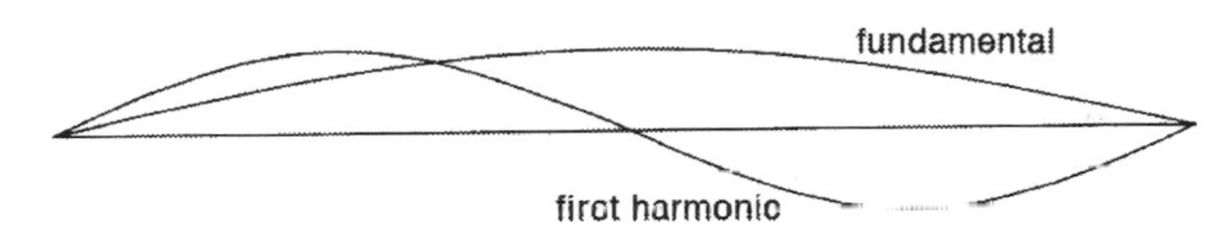

Figure 3.2. A snapshot of the fundamental vibration of a stretched string and its first harmonic.

as the *normal modes* of the string.) This is the scientific explanation of what Pythagoras was said to have conjectured more than two millennia ago without any notion of waves.

If the string is not fixed at one end but shaken with the frequency f while the other end is lose, there are traveling waves of the mathematical form $\sin(2\pi x/\lambda - 2\pi tf)$, of wavelength $\lambda = c_{st}/f$ whose maxima move with the fixed speed $c_{st} = f\lambda$.

So here are the important properties of solutions of the wave equation, an equation that we shall encounter again in other quite unrelated contexts: its solutions are oscillatory with an arbitrary amplitude and frequency f; their shapes in space are sinusoidal waves — sinusoidal means having the shape of the sine wave shown in Fig. 3.1 — with a wavelength $\lambda = c_{st}/f$, where c_{st} is the constant that appears in the wave equation; and all these waves travel with the fixed speed c_{st}. The amplitude and frequency of any given solution are determined not by the equation but by external circumstances called either boundary conditions or initial conditions. Furthermore, the sum of any two solutions of a given wave equation is also a solution — this is called the superposition principle. A plucked string does not necessarily oscillate with a single one of its proper vibration frequencies but usually with a superposition of several of them at the same time; these are the 'overtones.' Different instruments produce different admixtures of such overtones, which allow us to recognize their various timbres.

Remarkably enough, the French mathematician Jean Baptiste Joseph Fourier proved that any vibration of a given string, no matter how complicated its shape, can always be written as the superposition of its proper vibrations, each with its own amplitude, and if necessary, infinitely many of them. Born in Auxerre in 1768, Fourier was educated at the local military academy, at a Benedictine school, and at the École Normale (later called the École Normale Superieure) in Paris. After the Revolution, which he had the luck to survive, Napoleon appointed him to a variety of administrative positions, for which he showed such a great aptitude that he was made a count and later a baron. All the while, Fourier did mathematics in his spare time. When the emperor returned from his exile on Elba, Fourier resigned his administrative positions in protest against Napoleon's autocratic rule but somehow managed nevertheless to become head of the Bureau of Statistics, where he could do pure mathematics full time. He died in 1830 as a result of a disease he had contacted in Egypt during Napoleon's campaign there. Fourier's theorem, that any periodic function — any function that repeats itself *ad infinitum* — can be decomposed into a sum of trigonometric functions (the series of amplitudes of these sine and cosine function is called the *Fourier transform* of the original function) turned out to be of great utility in solving many linear differential equations: the equation satisfied by the transform was often easier to solve. Modern physics could not do without it. Sinusoidal waves are ubiquitous.

There are other musical instruments, such as drums, that produce oscillations in a manner very similar to string instruments but in two dimensions rather than one, and their motions are governed by the two-dimensional version of the wave equation. The proper frequencies of such membranes are determined, like those of strings, by the standing waves that are possible on them when they vibrate. In two dimensions, however, the spatial configurations of such standing waves, determined by the weight, tension, size, and shape of the membranes are much more complicated than in one.

A method of making the shapes of such standing waves visible was invented in the late eighteenth century by the German lawyer and amateur musician Ernst Chladni. He would use metal plates, usually clamped fixed in their center, sprinkle fine sand evenly on them and make them vibrate by stroking an edge by means of a violin bow. The oscillations would make the sand accumulate at the places that remained still, the nodes of the standing waves, making these nodes quite visible.[2] A drawing of the sand distributions Chladni found in one of his experiments is shown in Fig. 3.3. When Napoleon saw such a demonstration by Chladni, the emperor was so impressed that he not only paid him a handsome sum but offered a considerable prize to anyone who could explain these patterns. And so he did in 1816 to the French mathematician Sophie Germain. For plates of irregular shapes, the Chladni patterns remain not fully explained to this day.

Now the next question is of course: how do the vibrations produced by a musical instrument and by sound sources such as

[2]In the one-dimensional case of strings of length L, the nodes are the points at distances L/n, $n = 2, 3, 4, \ldots$ from one end.

Figure 3.3. A copy of a drawing of Chladni's original publication (from paulbourke.net/geometry/chladni/).

the human voice or explosions reach our ears? It took millennia for the nature of sound to be understood, as it was long thought to consist of a stream of particles and therefore would require no air for it to be heard. The German Jesuit scholar Athanasius Kircher in the seventeenth century was the first to devise a test to determine if air was indeed needed for sound to be heard. Born in 1601 in the small German town of Geisa, Kircher was a man of an enormous range of interests, and wrote some 40 works on subjects ranging from oriental studies to medicine and geology. The experiment he performed consisted of a ringing bell enclosed in an airtight jar from which he gradually removed the air by means of a pump, making the sound gradually dimmer. However, it took the much improved airpump of the English physicist and chemist Robert Boyle to make his experiment — which to this day is performed in elementary physics classes for the same purpose — really convincing by totally silencing the ringing bell: the propagation of sound indeed requires the presence of air.

The science of acoustics, the propagation of sound in air or other gases, begun in the context of music by Mersenne, was finally completed, more or less, by the English physicist John William Strutt in his treatise *The Theory of Sound*, the first volume of which was published in 1894.

One of the very few hereditary English peers who became prominent scientists, John William Strutt was born in 1842 in Langford Grove, Essex, attended Harrow School and obtained his Bachelor of Arts and Master of Arts degrees in mathematics at Trinity College, Cambridge, where he also became a Fellow. On the death of his father, the 2nd Baron Rayleigh, he became the 3rd Baron Rayleigh and has been generally known as a physicist under the name of Lord Rayleigh. He became Professor of Natural Philosophy at Cambridge and served as President of the Royal Society from 1905 to 1908. In 1904, the fourth year the prize was awarded, he won the Nobel Prize in physics. Lord Rayleigh died in 1919 at Terling Place, Essex.

The vibrating string produces oscillating pressure waves in the air with the same frequency as the string and governed by the wave equation too — the three-dimensional version of it, applicable to three-dimensional space rather than a one-dimensional string — but with its own constant c_s, the precise value of which depends on the temperature, pressure, and humidity of the air. Under normal conditions, the speed of sound c_s in the open air is about

343 meters/second.[3] The frequency of the middle C note being 261.6 Hz,[4] it follows that the wavelength of that sound in air is 1.31 meters.

You can use the sound-speed of 343 meters/second to calculate how far away a thunderstorm is by counting the time delay between the lightning you see and the accompanying thunder you hear. Since the speed of light is so enormous — approximately 300,000,000 meters/second — you may assume that the instant you see the lightning was also the instant at which the thunder was emitted, and a count of five seconds between your seeing the former and hearing the latter indicates an approximate distance of a mile.

Just as the waves on a string are small variations in the string's distance from its equilibrium position, so sound waves are very small deviations of the air pressure from its ambient value. Producing these pressure variations, the air molecules have to move slightly forward or backward from their normal average positions, which makes sound waves *longitudinal* rather than *transverse* waves like those of a string, the vibrations of which are at right angles to their propagation along the string.

The sound waves caused by the vibrations of the musical instrument or other source now move away, spreading with a spherical front at the speed c_s. As they move, their amplitude is gradually attenuated, partly because of the growing size of the spherical front — the total energy of the pressure wave needs to be conserved — and partly because the irregular longitudinal molecular motion making up the sound pressure keeps interacting with the ambient molecules of the medium.

---

[3]Note that the velocity with which a sound travels does not depend on how loud it is; the sound of an explosion travels no faster than a whisper.

[4]The unit of frequency is Hertz (Hz), which means oscillations per second. It is named after the German physicist Heinrich Hertz, who lived in the second half of the nineteenth century. We will meet him again later.

What happens when a sound wave encounters an obstacle such as a hard wall? Since the air molecules there are prevented from further motion, the solution of the wave equation at such a point has to have a node. This means that there has to be another wave of the same frequency so that its superposition with the incident wave — remember that solutions of the wave equation obey the superposition principle, which means the sum of any two solutions is also a valid solution — vanishes at the wall: this second wave is what we mean be the reflection of the first. If a sound wave strikes a wall making an angle α with the vertical, the reflected wave leaves the wall making the angle $-\alpha$ with it, and in the same plane, just like an elastic ball. When you are in a position to hear both the original sound and the reflected wave, you hear it as an echo. Even vast rooms like concert halls are rarely large enough for such delays between the original sound and its reflection to hear an echo; outdoor obstacles like mountains are required for that. If the obstacle is soft, such as a cloth curtain, the pressure wave is simply absorbed and the sound wave is reflected weakly or ends at that point. The result, in an enclosure, is a large conglomeration of waves. The overall effect of the combination of original musical sound and its, generally muted, reflections is that you hear any given sound dying away slowly: this is called *reverberation*. The reverberation times of various concert halls vary greatly and are the crucial criteria by which the quality of such halls is judged by musicians and audiences.

In a sense, these waves do not interact with or disturb one another. Nevertheless, when two of them with the same wavelength travel along the same direction along the same line, they produce a noticeable phenomenon called *interference*, the nature of which depends on their relative phase difference. What this means is the following.

Suppose that one of the waves traveling in the direction along the x-axis is of the form $A \sin(2\pi x/\lambda - 2\pi tf)$ while the other is of the form $B \sin(2\pi x/\lambda - 2\pi tf - p)$, where p is the phase difference. If $p = 0$, the two are said to be 'in phase.' This phase difference can

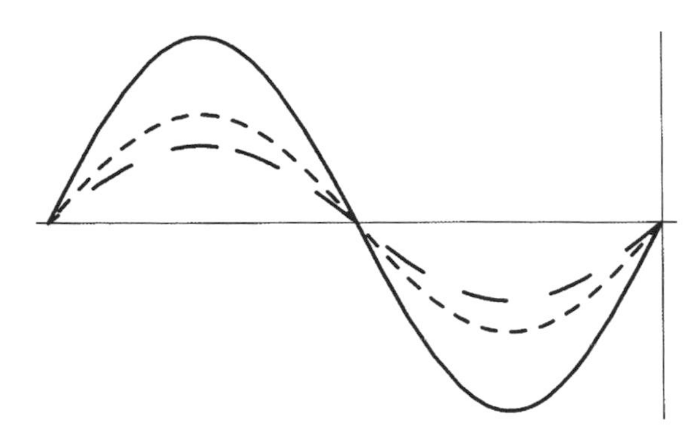

Figure 3.4. Two waves in constructive interference: the solid curve is the sum of the two dotted ones.

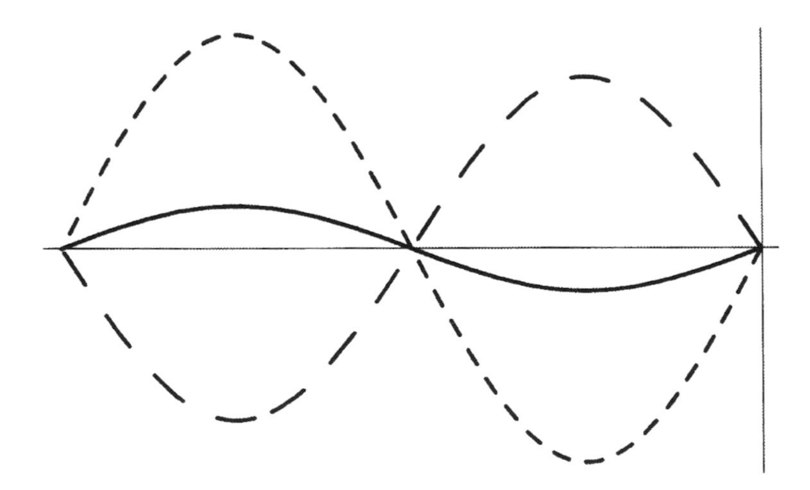

Figure 3.5. Destructive interference: again, the solid curve is the sum of the two dotted ones.

have very different effects on the sum of the two waves, as shown in Figs. 3.4 and 3.5. The first shows the superposition of two waves that are in phase with one another, and the other shows the superposition of two waves that are 180° out of phase.

There is, however, also another effect that adds to the complication of the sound quality of a concert hall, making the architecture of such halls an enormously complicated art. The superposition

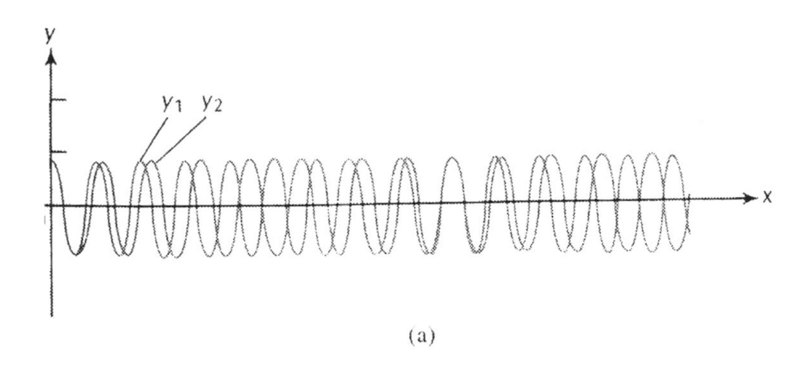

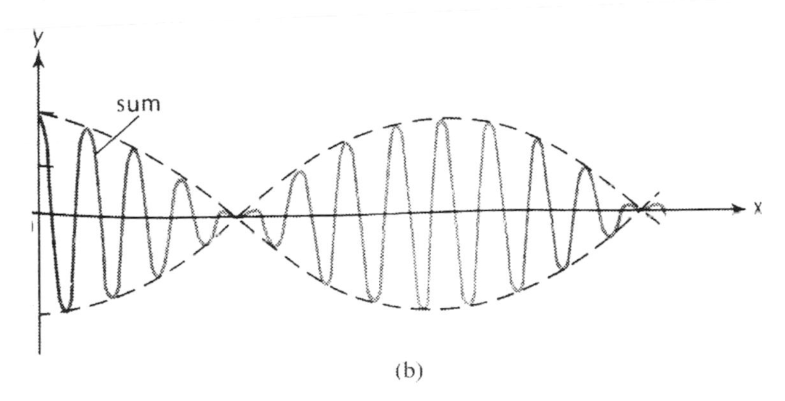

Figure 3.6. (a) shows two waves with slightly different wave lengths, and (b) shows their superposition: the resulting beats.

of two waves with slightly different frequencies, and hence slightly different wave lengths, is shown in Fig. 3.6: the loudness of the sound you hear slowly varies rhythmically, a phenomenon called *beats*. You can hear this effect by listening to two tuning forks or two musical instruments that are a bit out of tune.

As mentioned earlier, the speed with which sound travels in a given medium does not depend on its loudness. If the source is moving, however, its wavelength does depend on the speed and direction of the motion of the source. Since the advancing sound wave is caused by the periodic kicks that it receives from a vibrating source that is moving with the velocity v, the time from one crest of the wave to the next must equal the period T of the vibration of the source. Therefore, the distance between these crests must be $\lambda' = (c_s - v)T$ if the source and the sound are travelling in the same direction and $\lambda' = (c_s + v)T$ if they are travelling in the opposite direction. This shows that the wavelength of the sound emitted in the forward direction is shorter than that emitted in the backward direction. The frequency of the sound heard by a listener ahead of a moving source is thus higher than that heard by a listener behind it, and the faster the source moves, the bigger the difference: as a fast moving police car passes you, the pitch of the sound of its siren suddenly appears to drop. This easily observed phenomenon, entirely caused by the wave nature of sound, is called the *Doppler effect* after the Austrian physicist Johann Christian Doppler, who lived in the first half of the nineteenth century.

While the transmission of sound does require a medium, that medium need not be a gas like air. Sound waves also travel through liquids like water (about four times faster than in air) as well as through solids such as iron (about fifteen times faster than in air) or walls. Other properties of sound waves, such as their periodicity and their relation between wavelength and frequency, are the same in such mediums as in a gas; they are still governed by the wave equation. Their gradual attenuation, however, tends to be much larger in a liquid than in a gas.

The reflection of sound by a hard obstacle can be employed to locate it, as is done by the use of sonar, especially underwater. The use of such devices for submarine warfare has played an important role in the First World War and especially in the second. A ship would send out a short sound pulse — a 'ping' — and then listen for an echo produced by its reflection from an obstacle such as a submarine. Not only would the echo show the existence of a submarine, but the delay between the emission of the pulse and the reception of the echo would give information of its distance. The Doppler effect could even be used to measure its speed.

In order for our senses to register the presence of a sound wave, of course it has to reach our ears and we have to hear it. This happens only if their frequency is higher than about 16 Hz and lower than about 20,000 Hz. (As you get older this upper limit tends to decrease.) Sound with a higher frequency than that — implying a wavelength in air downward from about[5] 1.7 cm — is called *ultrasound* and, though we cannot hear it, it has a variety of practical uses. The vibrations causing it are produced primarily by electrical means called *transducers*, which are also employed to detect such sound.

One everyday use of ultrasound is for the analog of underwater sonar in automobiles. This allows your car to detect nearby obstacles and show their distance on a screen, making parking in tight spaces easier. More important applications are in medicine, where it can be used very similarly to x-rays in cavities of the body that are filled with liquid, without the latter's deleterious side effects. As the sound wave travels through the liquid, for example in a woman's uterus containing a fetus, its amplitude is attenuated depending on the density of any obstacle encountered, producing an image just like x-rays. Other medical applications of ultrasound include a non-invasive treatment for kidney stones called lithotripsy. Here ultrasound imaging is used to locate a kidney stone, and then a focused, high-intensity ultrasound pulse is applied to shatter it, the small fragments to be naturally eliminated without pain. The reflection of ultrasound waves is utilized to produce moving images of the heart in echocardiograms.

There was, however, also a quite distinct kind of sound wave, called a shock wave, first described by the nineteenth century Irish physicist George Gabriel Stokes, though its existence was doubted by many others at the time. Shock waves are produced when an object travels through the air faster than the speed of sound.

The ratio of the speed of a shock-wave producing object, such as a bullet or an airplane, to the speed of sound is called its *Mach number* in honor of an Austrian physicist who did research on such waves. In the late nineteenth century, Ernst Mach was professor of philosophy

[5]A centimeter, cm, is one hundredth of a meter.

Figure 3.7. A photograph of shock waves produced by a model of an X-15 aircraft in a wind tunnel at Mach 3.5.

at the University of Vienna and famous at the time for his espousal of the philosophy of positivism, which became very influential for a while. (He was also infamous, however, for vociferously opposing the notion that atoms existed. Mach argued that physicists and chemists had no business assuming the reality of atoms, tiny objects that no one had ever seen.)

These waves are of a completely different nature than ordinary sound waves and they are not governed by the wave equation. Rather than small variations in the pressure of the air like sound waves, they are very sharp, discontinuous changes in air pressure forming cones centered near points on the surface of the fast travelling object, and when they reach your ear, you hear them as sharp, loud cracks usually called sonic booms.

After our discussion of the sound produced by the vibrations of strings such as used by musical instruments you may get the impression that the waves associated with oscillations are always gentle and far removed from such horrid phenomena as tsunami. This impression would, however, be quite wrong. What happened on November 7, 1940, to the Tacoma Narrows Bridge in the state of Washington shows that even such simple oscillatory waves can be enormously destructive. There a wind blowing across the main span of the bridge produced eddies that began to set it to vibrate and procuce a standing wave — a bridge is an elastic elongated object, just like a string, only much stronger — that created a resonance.[6]

The roadbed of the main span of the bridge began to vibrate (see Fig. 3.8) increasingly violently until the strain became too great and it finally collapsed: the bridge was destroyed. (See Fig. 3.9.)

At this point we shall turn from the mechanical osillations and pressure waves to waves of a different nature, namely light.

Figure 3.8. The vibration of the Tacoma Narrows Bridge on November 7, 1940. (Image taken by Steven Rosenow of Shelton, Washington.)

---

[6]A resonance is a phenomenon when two vibrations are in phase and one reinforces the other, gradually increasing its amplitude. It's as though, instead of plucking a string on a guitar you made it vibrate by blowing at it rhythmically at the frequency of one of its normal modes, thereby slowly increasing its oscillation amplitude.

Figure 3.9. The collapse of the Tacoma Narrows Bridge on November 7. 1940. (Photo taken by Barney Elliott, The Camera Shop. Wikipedia.org).

4

Electromagnetic Waves

Let There be Light ...

Though the earlier Greeks thought that vision worked by means of rays sent out by our eyes, the nature of light had not been in serious contention since the later years of ancient Greek culture. Until the scientific revolution in the seventeenth century light was thought by many — including Isaac Newton — to consist of particles. What was primarily controversial was the question of the speed with which it traveled: most thought that the speed of light certainly had to be infinite. This included the great French philosopher René Descartes, for whom the infinity of the speed of light was a fundamental postulate. Science and its methods, however, intervened.

It was the Dutch astronomer Ole Christensen Römer who, in 1679, proved Descartes wrong and measured the velocity of light for the first time. His measurement was based on his observations of the variations in the times at which the planet Jupiter eclipsed its moon Io, depending on Jupiter's distance from the Earth. Based on his observations, he predicted a ten-minute delay of the next eclipse, and he explained the sensational success of his prediction on the finite speed of light. It would take light about 22 minutes to cross the Earth's orbit and 11 minutes to travel from the Sun to the Earth, he proclaimed. When the Dutch scientist Christiaan Huygens subsequently converted these numbers into velocity, they came out to be[1] 225,000 km/sec, a number that is remarkably close to the now accepted value of 299,792.458 km/sec.

The question of the nature of light, however, still remained unsettled. Among the ancient Greeks there were two opposing trains of thought. While the Pythagoreans held that light was emitted as a stream of particles, Aristotle saw it more like the waves on the ocean. Although research on optics had been pursued during the Middle Ages both by the Persians and the Arabs as well as to some extent in Christian Europe, a serious attack on the question of what light actually was had to wait until the scientific revolution of the seventeenth century. Christiaan Huygens was the first to develop a specific theory dealing with this question, and he followed more or less in the footsteps of Aristotle.

Born in The Hague in the Netherlands, the son of a diplomat, composer, and poet, Huygens had studied mathematics and law at the Universities of Leiden and Breda, but decided to devote himself to physical science. Performing experiments in a variety of fields, he left his most permanent imprint on optics. In a paper presented to the French Académie des sciences in 1678 and published as a book *Treaté de la lumière* in 1690, he proposed a theory that light was a wave that propagated in a material medium — similar to the way we now know sound works in air and in fluids (see Chapter 3) — which he called the ether.

Meanwhile, Isaac Newton had done pathbreaking experiments on optics too, discovering by means of a glass prism that the light we receive from the sun is actually a mixture of all the colors of the rainbow, and he was of the firm opinion — opposed by his usual adversary Robert Hooke — that light was made up of particles, which he published in his great treatise *Opticks* in 1704, not long after Huygens's book. Newton's fame was such that his opinion of course overshadowed the work of Huygens, but ultimately the question had

[1] 1 km (kilometer) equals 1000 meters.

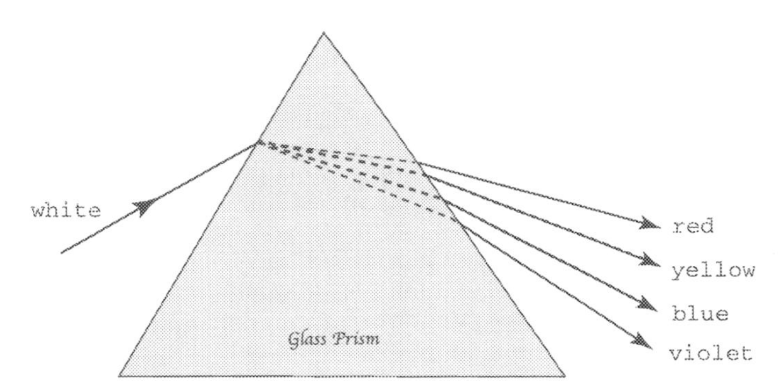

Figure 4.1. Newton's decomposition of white sunlight into its colored components by means of a prism.

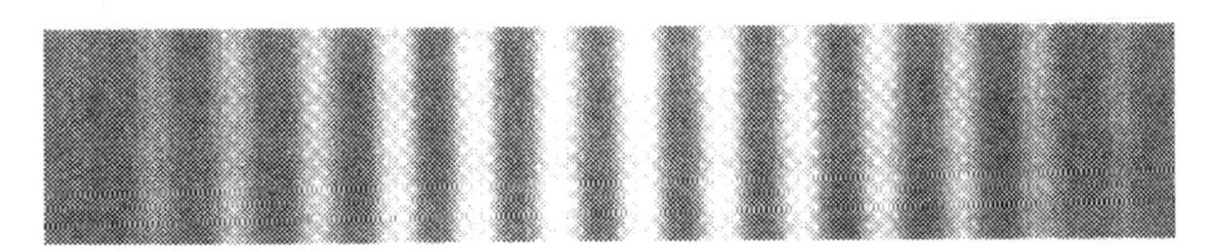

Figure 4.2. The image of a double-slit in an experiment like that performed by Thomas Young.

to be decided by experimental evidence, which was not provided until 1803 by the English scientist Thomas Young.

A polymath born into a Quaker family in 1773, Thomas Young became famous for contributing to the deciphering of the Rosetta Stone of Egyptian hieroglyphics. He also made notable contributions to physiology, language, musicology, solid mechanics, as well as other fields. In physics, however, he is best known for performing the crucial experiment that established the wave theory of light.

Convinced Huygens was correct, that light consisted of waves rather than particles, he demonstrated it by shining light through two slits and projecting the result on a screen. Instead of two bright strips, the images of the slits as they would appear if beams of particles had produced them, what the screen showed is reproduced in Fig. 4.2.

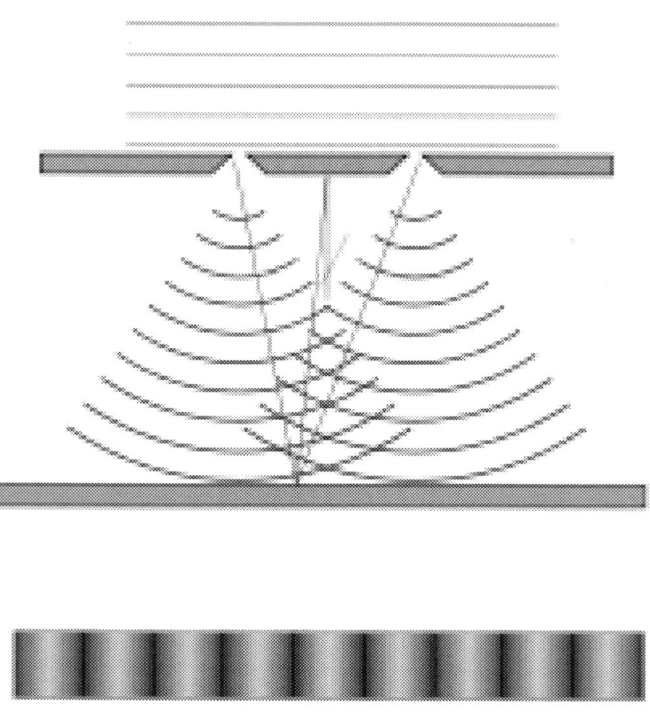

Figure 4.3. Diffraction and interference producing the image of a double-slit in the experiment of Thomas Young.

How did this come about? The explanation, shown in Fig. 4.3, rests on the phenomena of diffraction and interference, typical of the behavior of waves. As the light emerges from each slit, it flares, just as the water waves on the surface of a lake do after passing through an opening in a barrier: this is called *diffraction*. The emerging waves then interfere with one another (see Chapter 3), sometimes constructively and sometimes destructively, depending on their relative distances from the two slits. The result is what you see on the screen as multiple bright lines. There could no longer be any doubt that light consisted of waves rather than particles. The next question, then, was of course: waves of what? What was this 'ether' that Huygens had postulated and what kind of waves made up light?

Figure 4.4. This shows the refraction of light at a water surface.

Light waves have an important property that was discovered by the French engineer Étienne Louis Malus in 1808 by experimenting with and observing light shining through calcite crystals. When a light beam crosses from one medium to another at an angle, that angle usually changes, as you may notice when looking at an object partly under water (see Fig. 4.4 for an example). Such an object will seem distorted to you because the light reflected from the submerged part does not reach your eye in a straight line but is broken at the surface of the water. This effect is called *refraction*, and the amount of the angle change from vacuum is determined by the medium's index of refraction. (Newton's decomposition of sunlight into its differently colored components by means of a prism (see Fig. 4.1) rested on the fact that the index of refraction of glass varies with the color of the light, so that different colors are bent by different angles.) We shall return to this soon.

Now, there are certain crystals, called birefringent — calcite being one of them — that consist of layers with different indices of refraction. A monochromatic light beam passing through it emerges in two parts at different angles with different properties. When Malus shone these two emerging beams again on a calcite crystal, he found that, though of the same color, they behaved differently: they were differently polarized. (Nowadays it is quite easy to test the polarization of light by means of polaroid sun glasses.) It followed that light waves could not be longitudinal waves such as sound waves but had to be transverse, like the waves on a vibrating string. The polarization and the direction of propagation define a plane in which the wave oscillates at right angles to its forward motion.

The next step towards a full explanation of the nature of light was taken by Michael Faraday, whom we have already encountered in Chapter 2. Among Faraday's many disoveries was the fact that the polarization of light could be changed by a magnet. Clearly, therefore, light was an electromagnetic phenomenon. His conclusion was that light was simply a rapid oscillation of the electromagnetic field — a concept that he had invented himself — without the need for the ether postulated by Huygens.

However, the notion of an ether would nevertheless arise again anyway, with important consequences. When, near the end of the nineteenth century, all attempts to measure the velocity of the Earth relative to the ether failed, Einstein invented his counterintuitive theory of relativity (the so-called special theory) to explain it all: there was no ether; as far as light was concerned, no matter the observer's motion, the speed of light measured would always be the same. That's how the ether was finally put to rest by means of a new theory.

As we saw in Chapter 2, it was James Clerk Maxwell who provided the mathematical formulation of the precise properties of the electromagnetic field envisaged by Faraday. Maxwell's equations contained but one constant, usually called c, which had been found to be the velocity of light. Among the solutions of these equations in free space there are rapidly transversely oscillating electromagnetic waves traveling at the speed c. They have the same sinusoidal shapes and properties as the solutions of the wave equation described earlier, their frequency f related to their wavelength λ by the equation $f = c/\lambda$. In the frequency range visible to our eyes, the electromagnetic waves with different wavelengths are perceived by us as different colors, those with shortest wavelengths as violet-blue and those with the longest as red.

Maxwell's equations not only predict the existence of such waves and explain their nature, but they also imply that any accelerating electric charge — any charge that is neither at rest nor simply moving in a straight line at a fixed speed — necessarily emits radiation that at large distances can be decomposed into a superposition of them. These waves have many of the same properties as sound waves decribed in Chapter 3. They are reflected in a similar way by interfaces between materials of different optical properties, most effectively by mirrors, usually made of metal-coated glass. If the mirror is parabolic, the light from a point-source can be converted into a parallel beam, as shown in Fig. 4.5.

Another method of channelling light in specific directions is via optical fibers, very thin hollow tubes made of reflective glass

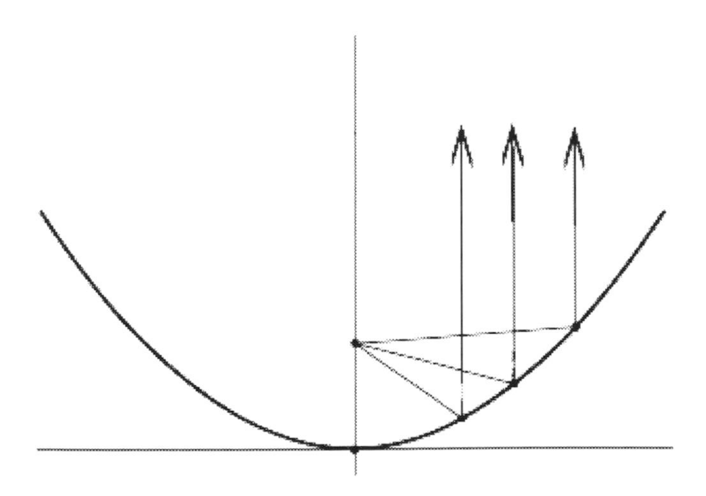

Figure 4.5. Reflection of light by a parabolic mirror from its focal point.

or plastic. Such 'light pipes' are extremely useful for illuminating or viewing hard-to-reach confined places, for example, inside the human body.

While the speed of light does not depend on the speed of its source of emission, it does vary from one transparent medium to another: the velocity of light in water differs from that in glass, etc., and it also depends somewhat on its color (an effect that is called *dispersion*). These speeds are always less than the speed of light in vacuum. To measure the enormous speed of light in the laboratory, as opposed to measuring it astronomically as Römer had done, is extremely difficult and was not accomplished until the nineteenth century by the French physicist Armand Hippolyte Louis Fizeau. His finding that light moved more slowly in water than in vacuum or air was another confirmation that it consisted of waves rather than particles. Newton, on the basis of his particulate theory, had predicted that it would move faster in water.

Variations in speed are the cause of different indices of refraction — the index of refraction of a transparent medium equals the speed of light in vacuum divided by that in the medium — and of the refraction of light at the interface between different media. The emerging ray lies in the same plane determined by the normal to the interface and the incoming ray, but their angles with the normal differ (see Fig. 4.6). Snell's law allows you to calculate very simply this change of angle: $n_1 \sin\theta_1 = n_2 \sin\theta_2$. Light traveling through a

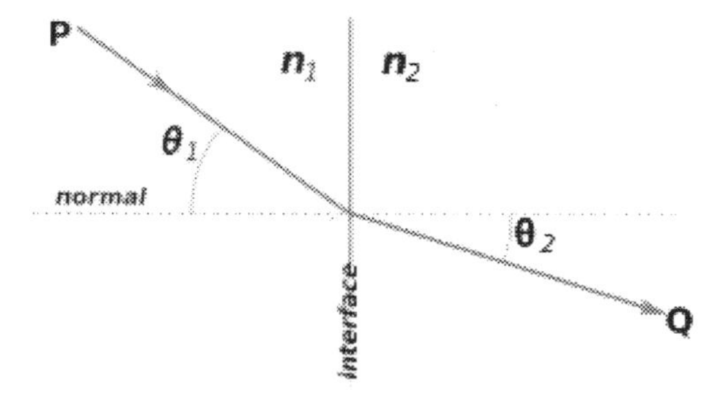

Figure 4.6. Refraction at the interface between two media of different velocities of light and hence different indices of refraction n_1 and n_2.

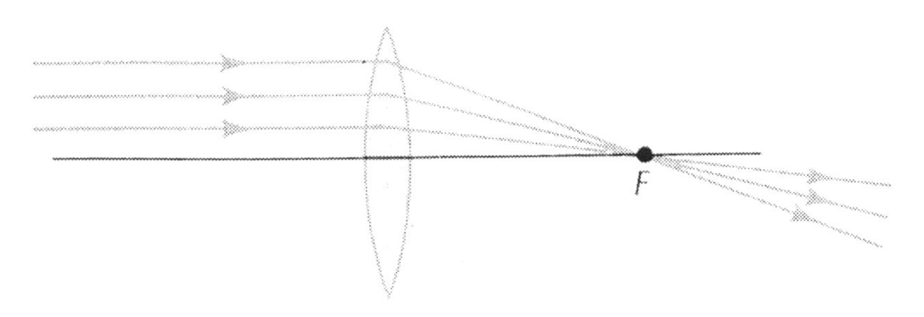

Figure 4.7. The focusing action of a convex lens. F is the focal point.

medium with a gradually changing index of refraction, as is the case for air at various temperatures, for example, will be gradually bent.

Refraction is the basis of the geometrical optics of lenses. A piece of glass with appropriately curved surfaces will have a variety of effects on transmitted light, depending on the surface curvatures. Figure 4.7 shows, as an example, how a convex lens focuses a beam of light. The enlarging properties of microscopes as well as those of telescopes are all due to refractive lenses, as well as to the reflecting properties of curved mirrors. In this context, dispersion is an undesirable side effect called chromatic aberration: different colors have different focal points. It can be counteracted by manufacturing lenses in layers made of glasses with different dispersions.

There is an interesting way of understanding why refraction occurs, called *Fermat's principle*: a light beam from point P to point Q will always choose the path that takes the least time. When point Q, the position of your eye, is in air and P in water, where the beam moves more slowly, for example, it would take more time to go straight from P to Q than choosing a somewhat longer path in air and a shorter one in the water, thus necessitating an angle change at the interface. Fermat's principle can be shown to lead directly to Snell's law. It's as though a light beam had a will of its own, seeking the fastest way.

Refraction has a variety of important consequences in the heavens. Why is the color of the sky blue, whereas astronauts report that in outer space it is black? The reason is that the light of the Sun is scattered by the small air molecules in the upper atmosphere, which have an index of refraction different from 1 and thus deflect or scatter light rays. For such very small obstacles — small relative to the wavelength of the light — the scattering probability, which physicists call the scattering cross-section, is proportional to the inverse fourth power of the wavelength. (First calculated by Lord Rayleigh, this is called Rayleigh scattering.) Thus blue light will be much more scattered, reaching our eyes from all directions of the sky, than the other colors, which have longer wavelengths.[2] As a result we see the entire sky as blue. In addition, the sunlight deflected by the air molecules is polarized, and most strongly so at right angles to the Sun. The long wavelengths are polarized the strongest of all. As a result, when observed through a properly oriented polaroid lens that filters out the strongly polarized rays, the sky at right angles to the sun appears more deeply blue, an effect sometimes exploited in color photograpy.

Rainbows, on the other hand, are caused by water droplets in the air which are large compared to the wavelength of visible light. A ray of light impinging on such a droplet at a given angle and surface position will be refracted. When it hits the surface again from the inside some of it will be refracted and escape, but a part of it will be reflected and remain inside. That part will hit the surface again, be refracted and emerge at another escape angle that depends on the refractive index of the water as well as on the position of incidence. If a broad beam of light shines on the droplet, the emerging rays will exit at somewhat different angles, and they will be focused at a point at a specific angle that depends on the refractive index, the droplet acting like a lens. As a result, if you are positioned with your back to the sun, your eye will receive the most intense refracted light from the droplets forming a large circular arc whose radius depends on

[2]Although the wavelength of violet light is smaller than blue, there is not as much of it in sunlight.

Figure 4.8. A photograph of a rainbow. (From scriptoriumblogorium.blogspot.com.)

the refractive index of the water droplets. Since that index depends somewhat on the color of the light and sunlight consists of 'all the colors of the rainbow,' these will be separated in the actual rainbow you see, as in Fig. 4.8. The light that emerges after having been once internally reflected will similarly emerge at a larger angle, sometimes visible as a somewhat fainter secondary rainbow in which the order of the colors is reversed. (The area between the two rainbows, usually a little darker than the rest of the sky, is called Alexander's band, named after Alexander of Aphrodisias, a Greek philosopher who lived near the end of the second century CE and was the first describe it.)

A more rarely seen phenomenon is caused by the strong scattering of light by the water droplets that make up clouds at an angle almost exactly in the backward direction. This results in a bright ring of light, visible for instance from an airplane, surrounding its shadow, as seen in Fig. 4.9. Since it was first observed by mountaineers standing on top of a mountain, with the sun behind them, seeing the shadow of their head surrounded by a bright halo, just as holy figures were depicted in medieval paintings, this phenomenon is called the *glory*.

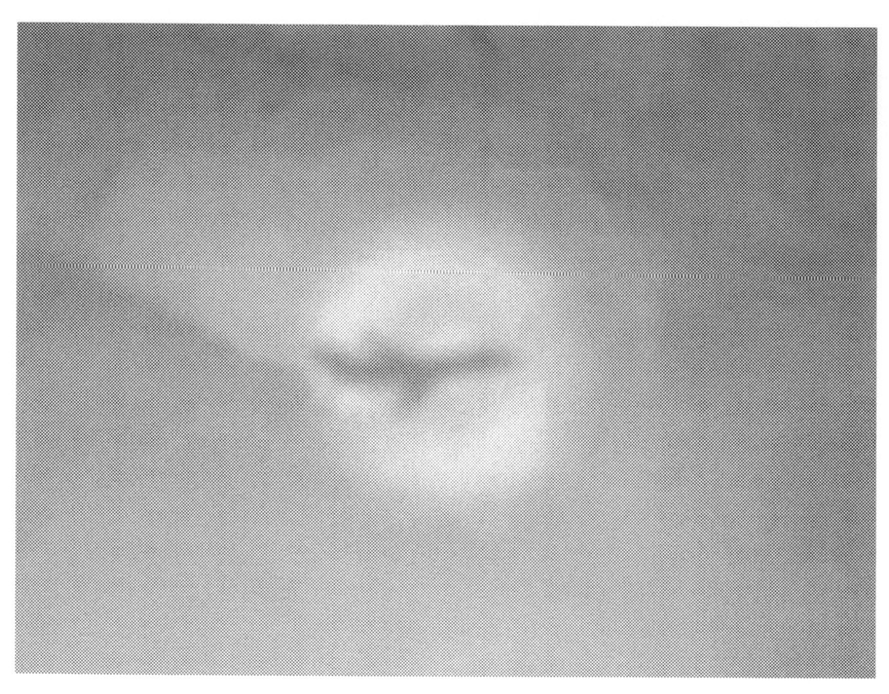

Figure 4.9. Solar glory photographed over Canada, August, 2005. (Source: http://home.comcast.net/milazinkova/Fogshadow.html Mila Zinkova).

Light waves are also subject to the Doppler effect: the wavelength of light emitted by an object traveling away from us is seen as longer, that is, shifted toward the red, whereas the light we see emitted by an object traveling towards us seems more blue. Imperceptibly small unless the emitting object travels relative to us with a speed comparable to the enormous speed of light, we are never aware of the Doppler shift of light in everyday life.

There was, however, an occasion when it played an important scientific role. In 1927 the astronomer Edwin Hubble at the Mount Wilson Observatory made the remarkable discovery that the spectra of the light emitted by far-away stars and galaxies all seemed to be shifted toward the red. As chemists had known for a long time, each element when heated emitts light of a characteristic set of frequencies called its spectrum, or specifically its emission spectrum

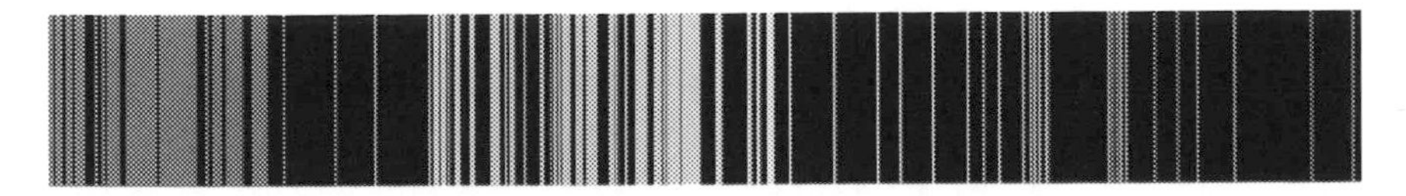

Figure 4.10. The emission spectrum of iron.

(see Fig. 4.10 for an example), identifying each element like a fingerprint. The light emitted by stars had been found to be composed of the same spectra as found on Earth, showing that the stars were made up of the same chemical elements as we know here. (The only exception was a spectrum contained in the light from the Sun that had never been seen on earth. It was later found to be that of a noble gas that was thereafter called helium in honor of the Greek god of the Sun.) What Hubble discovered was that very distant stars showed the same identifiable spectra, but the entire set of spectral lines was shifted toward the red. What is more, the further away the star was the larger was this red shift. There was much speculation about the cause of this spectral shift, but the only proposed cause that held up under close scrutiny was the one Hubble insisted on: it was a Doppler shift. This, of course, implied that the far-away stars and galaxies are all moving away from us, and the farther away the faster. In other words: the universe as a whole is constantly expanding like a bubble. And it was the acoustically well-known Doppler shift of the light waves that produced the signal of this expansion.

A few years after Maxwell's death, the German physicist Heinrich Hertz — as already mentioned earlier, the unit of frequency is now called hertz — discovered by means of experiments he performed, that the spectrum of electromagnetic waves predicted by the equations extends far beyond the visible. The wavelengths and the modern names of the various ranges are as follows: the visible spectrum ranges from about[3] 400 nm (violet–blue), green, yellow, and orange to 700 nm (red). Beyond the violet there are

[3] 1 nm (nanometer) equals 10^{-9} m.

the shorter and shorter waves called ultraviolet, X-rays, and gamma rays, as short as 10^{-16} m; the longer waves beyond red are called infrared, microwaves, then radio waves (the distinction between radio waves and microwaves is not really sharp, and they often lumped together) — with wavelengths from a few centimeters to thousands of meters. Those longer than that are simply referred to as long or electric waves.

Before discussing the interesting microwaves and radio waves, let us look for a moment again many years ahead. We now know that the laws of optics as normally formulated, especially those relating to the focusing of light beams, apply to natural light such as we receive from the sun or from light bulbs and such. They are necessarily subject to diffraction and tend to spread. Monochromatic light of this kind, as we now know, is produced by enormous numbers of atoms and it consists of light with different, totally uncorrelated phases (see Chapter 3). These cannot produce regular interference effects, which are the result of the superposition of waves with a fixed phase difference. In the early twentieth century Albert Einstein predicted the stimulated emission by atoms of light with different properties, and his prediction was verified some fifty years later as a *laser*, which is an acronym for **L**ight **A**mplification by **S**timulated **E**mission of **R**adiation. In a beam of such light all the waves have a fixed phase difference; they are marching in step, so to speak, and as a result the beam can be focused arbitrarily sharply and remains so without limit. Such highly focused laser beams have many important applications, such as being able to measure the distance from the Earth to the Moon within inches. This was done by means of reflectors that were left on the moon during the Apollo program. Measurement of the time it took for signals sent by a laser beam to be reflected by them and return allowed an easy and very accurate calculation of the distance.

The same principle that underlies the laser for light, of course also applies to microwaves. Its use there is called *maser*.

We shall now turn to microwaves and radio waves, and then specifically to the radio.

...Microwaves and Radio Waves

Just as light can be channelled in specified directions by optical fibers, so can microwaves by the use of waveguides. These are hollow tubes usually of rectangular cross-section, made of metal, that guide electromagnetic waves by repeatedly reflecting them from their walls. (Optical fibers are really nothing but waveguides specifically designed for light.)

One of the important uses of microwaves is in the well-known radar, an acronym that stands for **RA**dio **D**etection **A**nd **R**anging (it uses both radio waves and microwaves). Its principle is very simple: a microwave beam is directed by a large emission antenna at an object made of metal that therefore reflects electromagnetic waves, and the reflected beam is detected by an antenna usually located in the same place as the sending one. An antenna is simply a metal object that generates an electromagnetic field in the air when an oscillating current goes through it; when struck by such a field from the outside, it in turn generates a current that can be detected. It produces a beam by means of a large hyperbolically shaped metal surface in a way that is quite analogous to the way a hyperbolic mirror produces a beam of light as shown in Fig. 4.5. Pulsed radar, which sends short pulses and measures the time between the emission of a pulse and the reception of its reflection, gives additional information about the distance of the reflecting object.

Radar was invented by the German engineer and entrepreneur Christian Hülsmeyer, born in 1881 in Eydelstedt in Lower Saxony. He received a patent on his Telemobiloscope in 1904, intending its use for preventing collisions between ships in foggy weather. His invention was more or less forgotten until shortly before the Second World War, when both Britain and Germany began developing it further. While the German government did not regard it as important because it had no great offensive capabilities, the Royal Air Force made great efforts to develop it, and especially to make RDF (an acronym for **R**ange and **D**irection **F**inding), as they called it, small enough to be carried by fighter planes so as to detect German bombers in the air and not just from ground stations. The cavity magnetron, invented by John Randall and Harry Boot of Birmingham University in 1940, is a small device that generated microwaves extremely efficiently, and enabled the British to use RDF with great effect in its air defence. Further technical developments of radar during the war came primarily from the Radiation Laboratory (called the Rad Lab) at MIT, which at one time employed as many as 4000 people. Just as underwater sonar saved the Allies from the German submarines, so the Battle of Britain could not have been won by the British without radar. Today, radar is used extensively in air traffic control at airports all over the world. It is also used widely for weather forecasting, as microwaves are reflected not only by metal objects but also by the electrically conducting water in rain and thunderclouds. (Although water conducts electricity, it does so not nearly as well as metallic conductors.)

A quite different, smaller-scale application of microwaves takes place in the kitchen: the microwave oven. Confined to the interior of the metallic little oven, the microwaves heat anything containing water. This is because of the fact that water molecules are electric dipoles, and the rapidly oscillating mocrowaves' action on these dipoles makes them oscillate just as rapidly. This molecular motion is the essence of heat. The more water there is in the food item in the oven, the more it is heated, with the result that often its interior gets hotter than outside layers. (Of course, anything wrapped in metal foils is shielded from the microwaves and will not get heated.) This heating effect of microwaves was accidentally discovered in 1945 by Percy Spencer, an American engineer working at the company Raytheon. While he was working with radar, he noticed that the chocolate bar in his pocket began to melt, a misfortune for his clothing and his appetite, but good for the rest of the world. So Raytheon was awarded a U.S. patent and the first microwave oven went on sale in 1947.

Now moving from the kitchen to the universe, we shall describe the use of microwaves for astronomy. Astronomy has been practiced for millennia in all known cultures. The starry heavens and their varying configations were simply too fascinating to ignore. The invention of the telescope in the early seventeenth century made the science enormously more accurate, and astronomers used it very effectively to discover an ever-increasing number of properties of the universe, including its vast size, by observing its visible aspects. In 1933, however, the American physicist Karl Guthe Jansky serendipitously discovered that there were also sources out there that emitted invisible radio waves.

Born in 1905 in Norman, Oklahoma (still a Territory at that time rather than a state), his father the Dean of the College of Engineering at the University of Oklahoma at Norman, Jansky attended the University of Wisconsin at Madison, where he received his BS in physics. In 1928 he joined the Bell Telephone Laboratories at Holmdel, New Jersey, which was studying atmospheric and ionospheric properties by means of radio waves and microwaves, and his job was to investigate the sources of static that interfered with radio transmissions. Using a rotating antenna for the reception of radio waves of 20.5 MHz[4] (which have a wavelength of about 14.6 m) he recorded static from thunderstorms at various distances, but he also detected a strange kind of hiss, the intensity of which peaked once a day, with a period of exactly 23 hours and 56 minutes. This led him to conclude that the source of this radiation was not the Sun, which should lead to a period of 24 hours, but the Milky Way, a conglomeration of billions of stars. Unfortunately, Bell Laboratories denied him funding for a larger antenna that would have allowed him to pursue his finding of radio emissions further. It was 1933, the middle of the Great Depression, and no observatory had the money to follow up the discovery, so it remained relatively dormant for years.

In 1964, however, the use of radio astronomy led to a fundamental discovery about the universe, again at Bell Laboratories. Arno Allan Penzias was born in 1933 in Munich, Germany, the son of a Jewish

[4]MHz stands for megahertz, one million Hz.

businessman and a Jewish mother, who managed to get him and his brother into the *Kindertransport* to England at the age of six to escape Nazi persecution. His father and mother followed soon thereafter. Shortly after the outbreak of the war, the family arrived in New York and Arno began going to elementary school and on to City College, where he fell in love with physics. After graduating and spending two years in the U.S. Army Signal Corps, he received his Ph.D. in 1961 at Columbia University by building a maser amplifier for a radio-astronomy experiment. He then joined Bell Laboratories, as did another young radio astronomer, Robert Woodrow Wilson.

Wilson, born in 1936 in Houston, Texas, had studied physics at Rice University and the California Institute of Technology. He and Penzias built a supersensitive antenna to detect very faint radio waves bounced off echo balloon satellites. In order to succeed they were careful to eliminate all sources of interference they knew about, including the effects of radio broadcasting and radar, and even suppressed interference from heat in the receiver by cooling it with liquid helium. However, in spite of all their precautions, they detected a constant mysterious, steady noise with a wavelength of 7.35 cm, night and day, evenly from all directions in the sky, not emanating from the Sun or our galaxy. They knew of no radio source that could be blamed for this noise, nor could some pigeons or their droppings on the antenna, which they cleaned. What could account for this strange ubiquitous radio noise?

As they discovered, there had been two scientific articles in 1948, one by George Gamow and his student Ralph Alpher, followed by another by Alpher and Robert Herman, which indicated that they had discovered something quite fundamental.

George Gamow, born in 1904 in Odessa, Russia, educated at Novorossiya University in Odessa and the University of Leningrad, was a theoretical physicist and cosmologist who had learned about and worked on the new quantum theory at the Universities of Göttingen and Copenhagen from 1928 to 1931. He then obtained a position in the Radium Institute in Leningrad. However, when

Figure 4.11. The Horn antenna used by Penzias and Wilson to discover the cosmic microwave background radiation.

political repression in the Soviet Union became unbearable, he and his wife, also a physicist whom he had recently married, made two very venturous but unsuccessful attempts at escaping from his homeland. In 1933, however, they were both allowed to attend a physics conference in Brussels, and they never returned. The following year they moved on to the United States, where he became a Professor of Physics at George Washington University and then at the University of Colorado at Boulder, where he died in 1968.

Gamow's 1948 paper with Alpher (actually, it was published under the names of Alpher, Bethe, and Gamow, without a contribution by Hans Bethe nor his permission, because Gamow could not resist the pun — and it has ever since been known as the[5] α, β, γ paper) strongly supported the then quite controversial 'big bang' theory of the universe. Furthermore, it argued that all the chemical elements were gradually formed from the initial constituents of the universe, mostly neutrons, protons, and very intense electromagnetic

[5]α, β, γ are the first three letters of the Greek alphabet.

radiation, a process that became known as nucleosynthesis. If very high energy radiation had been left over from this process, it would have destroyed the newly formed nuclei. Therefore there had to be an upper limit to the leftover radiation. The paper by Alpher and Herman calculated, on the basis of the known remaining hydrogen in the universe — protons are nothing but hydrogen nuclei — and taking into account the red shift caused by the expansion of the universe, that the peak of the leftover radiation should be the equivalent of the natural radiation in a cavity (this cavity being the universe as a whole) with a temperature of about 5 K.[6]

The radio noise that Penzias and Wilson observed at a wavelength of 7.35 cm corresponded to a cavity temperature of 3.5 K. So it was clear that what they had heard was, in a manner of speaking, the crying of the infant universe. They had made a fundamental discovery by means of radio astronomy.

...and the Radio

Although the first practical use of the electromagnetic waves discovered by Heinrich Hertz was for wireless telegraphy, radio waves are of particular interest, and we will want to understand how they manage to transmit sounds — and in television, even pictures — over long distances. For this we will have to undertake a little detour.

The basic elements of an electric circuit, called an LC oscillator, are an inductor L and a capacitor C. A capacitor is a device in a circuit consisting of two flat parallel metal plates at a close distance from one another, separated either by air or a dielectric. When one of the plates contains negative electric charges (excess electrons) the other will contain positive ones (a shortage of electrons) because the electrons there are repelled by the electrostatic repulsion between like charges. If left this way, the charged capacitor contains a certain

---

[6]K denotes the Kelvin temperature, whose degrees are the same as centigrade but which starts at −273.16° Celsius.

amount of energy residing in the electric field between the two plates. Its capacitance C is proportional to the size of its plates and depends on the distance between them as well as on the dielectric material between them. The inductor consists of a long, tightly wound helical coil of wire, and its inductance L depends on the size and length of the coil and on the material about which it is wound. When a current flows through the wire, Ampère's law, mentioned earlier, dictates that it induces a magnetic field at the center of the coil.

A current in a wire circuit containing both a capacitor of capacitance C and an inductor of inductance L — an LC oscillator — will vary sinusoidally at a characteristic proper frequency $f = 1/\sqrt{LC}$, with its energy contained in the electric field of the capacitor when it is fully charged, and in the magnetic field of the inductor when the full current flows through it. To initiate the oscillation, the circuit is connected to an external sourse of energy, the strength of which determines the amplitude of the resulting oscillation. Since every circuit (unless at extremely low temperature) also contains resistance in its wiring, which converts some of the electrical energy of the current into heat, the oscillation is damped and will gradually disappear unless maintained by the external source.

Finally, the oscillator is connected, via a transformer, to an antenna, which usually consists of two metal rods or thin wires.[7] The resulting electric charges in the antenna, varying with the rhythm of the LC oscillator, produce an electromagnetic wave field in the surrounding space that will now travel away, broadcast in all directions with the speed c of light. Its frequency, called the carrier frequency, will be that of the originating LC oscillator, with a slowly varying amplitude determined by its energy source. If the strength of that source varies according to a sound source, the resulting electromagnetic waves sent into space will look something

[7] A transformer is a device that takes advantage of both Ampère's law — a magnetic field produced by the current in a wire coil in the LC oscillator — and Faraday's law — a current produced in another nearby coil, immersed in this field.

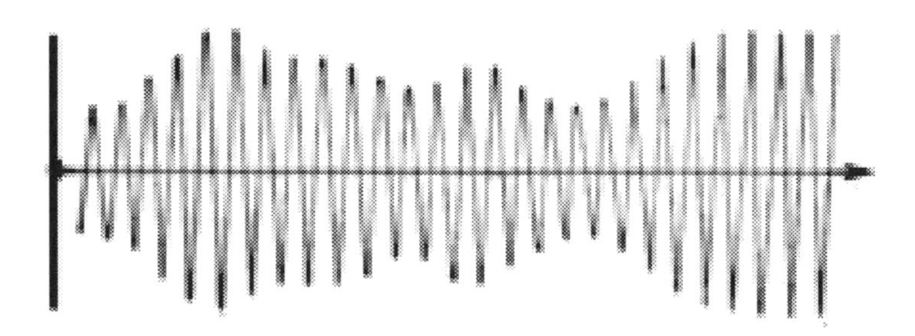

Figure 4.12. A radio carrier wave modulated by a sound wave.

like that shown in Fig. 4.12. (Remember that the frequency of the electromagnetic carrier wave is much larger and its wavelength much smaller than those of sound waves modulating it.)

Space being filled with electromagnetic waves of a great many different carrier frequencies — every radio station has its own — it is now the task of each individual radio to pick out its choice and convert the received signal into the sounds sent by the chosen station. How is that done? The procedure used by the most primitive radio sets show it in its purest form.

In order to pick out the broadcast of a specific station, the first requirement is for the receiver to have an antenna coupled to an LC circuit tuned to the carrier frequency of that station. It is easy enough to vary the proper frequency of an LC circuit by changing its capacitance: if one of the plates of the capacitor is hinged and can be rotated parallel to itself so that the area facing the other plate varies, the capacitance C changes and so does the proper frequency of the LC circuit. The properly tuned circuit will now be set to oscillate in resonance if the antenna is struck by electromagnetic waves of the chosen frequency, as shown in (a) of Fig. 4.13, duplicating that sent, Fig. 4.12.

In this form the received signal cannot be converted into sound by an earphone, because its negative voltage always equals the positive in short succession and there is no net force on its diaphragm. So the next step, the crucial one, is to employ a rectifier, a device that

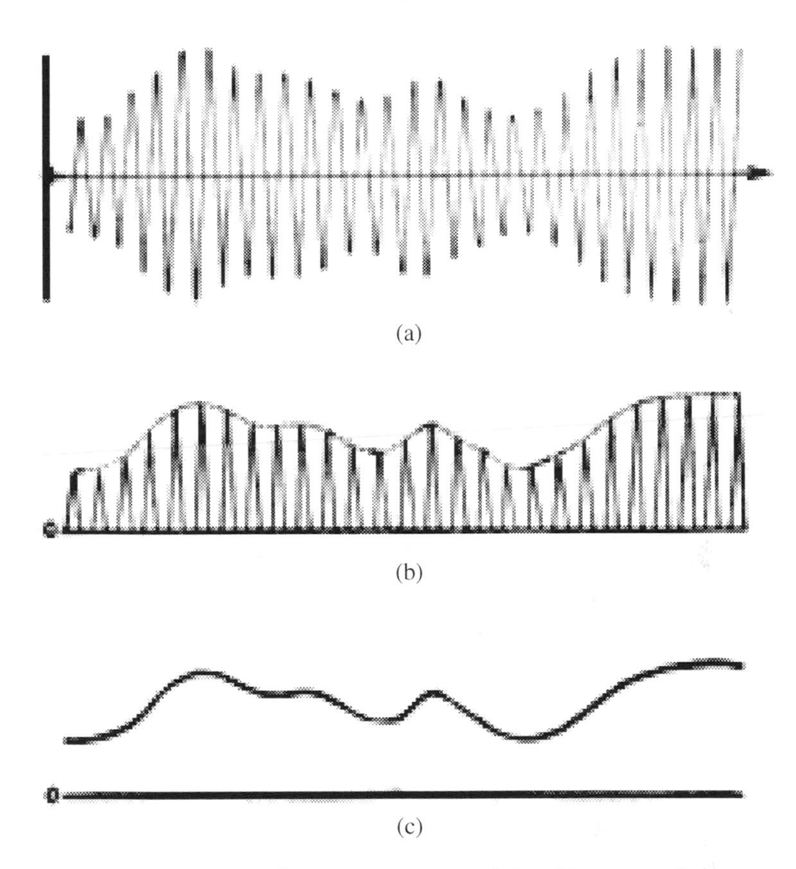

Figure 4.13. A radio carrier wave modulated by a sound wave.

converts alternating into direct current by allowing current to pass through it in one direction only. This is accomplished by the 'cat's whisker,' an early kind of crystal rectifier, subsequently called a crystal diode (see Fig. 4.14).

The 'cat's whisker' consists of a thin elastic piece of wire whose tip very lightly touches a crystal semiconducting mineral, for example galena, forming what would later be called a point-contact metal–semiconductor junction, or a diode. The current in the receiver is thus changed from the form (a) in Fig. 4.13 to the form (b). Finally, an element in the LC oscillator circuit of the receiver filters out the high-frequency part of the current so that

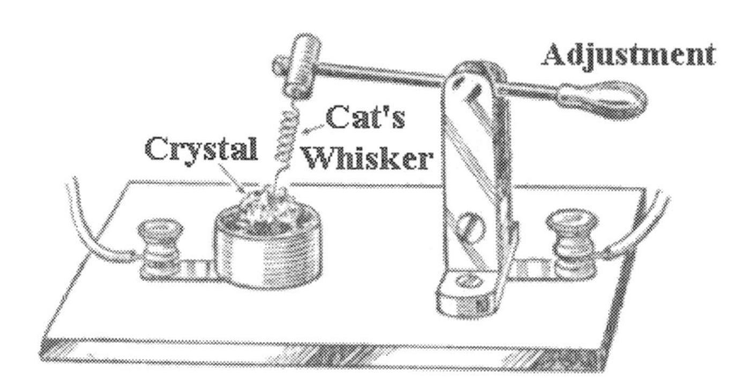

Figure 4.14. A drawing of a cat's whisker diode (from ipkitten.blogspot.com).

the remaining current looks like (c) in Fig. 4.13, the audio part of the received signal, appropriate for earphones. A simple radio of this kind did not require a power source, not even a battery. (I remember using just such a home-made crystal radio with a 'cat's whisker' as a teenager in 1938 listening, with great excitement, to the famous world heavyweight championship boxing rematch between Joe Louis and Max Schmeling.)

Any modern radio (and television too) works on the same principles, except that the devices implementing each of the necessary steps are much more efficient and work more reliably though, of course, they require an external power source. The crystal diode was replaced by a vacuum tube and later by a tiny semiconductor diode. Furthermore, an amplifier is added at the end that converts step (c) in Fig. 4.13 into a much stronger signal suitable for driving a loudspeaker rather than just earphones.

So now that we understand a variety of electromagnetic waves, both those of light and the longer radio waves, we'll turn to quite different kinds of waves, some of them often referred to as matter waves.

5

Wave Mechanics and Gravity Waves

In this chapter we shall be discussing two quite different wave phenomena, each governed by one of the new physical theories promulgated during the twentieth century: the first from quantum theory, the second from the general theory of relativity.

Wave Mechanics

The year 1905 was magic for Albert Einstein, the young clerk working in the patent office in Bern. In one of the three path-breaking papers he wrote that year — the one usually referred to as explaining the photoelectric effect and for which he was eventually awarded the Nobel Prize — he initiated the quantum theory of radiation. He postulated that electromagnetic radiation of frequency f, including light, always came in the form of 'quanta,' later called 'photons,' with the energy $E = hf$, where h is a universal, fixed number originally introduced by the German physicist Max Planck and now called 'Planck's constant.'[1] Does this mean that Einstein denied the validity

[1] For a thumbnail biographical sketch of Max Planck see my book *Why Science?* World Scientific Publishing Co., 2012. The book by J. L. Heilbron, *The Dilemmas of an Upright Man: Max Planck as Spokesman for German Science*, University of California Press, 1986, provides much more detail.

of Young's crucial experiment? Not at all. His idea introduced one of the big puzzles of the quantum theory: light of a given color is both a wave of a certain frequency f *and* also made up of a kind of 'particles,' photons. If this duality seems weird and hard to understand — too bad; that's the way nature is.[2]

Some nineteen years later, a young doctoral student by the name of Louis-Victor de Broglie was writing his dissertation in physics at the Sorbonne in Paris, and it occurred to him that Einstein's peculiar notion of light being both a wave and a particle may be a universal phenomenon. Perhaps all particles, such as electrons, were also at the same time waves, and he even postulated that if a particle had the momentum[3] p, then the wavelength of the associated wave should be $\lambda = h/p$. If that were so, it might even 'explain,' in a manner of speaking, the revolutionary model of the atom that had been introduced with great success by the young Danish physicist Niels Bohr not long after, though quite independent of, Einstein's quanta of light. Bohr's atom had electrons circulating about a heavy nucleus in orbits with only certain specific 'quantized' energies. Perhaps the lengths of those orbits needed to be exact multiples of the wavelengths of the orbiting electrons so as to form standing waves?

His bewildered thesis supervisors consulted Bohr in Copenhagen, who was skeptical, and Einstein in Berlin, who was enthusiastic.

Louis-Victor de Broglie did get his doctorate. Not only that, but it did not take very long for his idea to be experimentally verified in two different laboratories: by two American physicists Clinton Joseph Davisson and his assistant Lester Germer, and almost simultaneously

---

[2]Niels Bohr would later insist that the correct way of putting it was: that's the language in which we have to describe nature. Reality — the way nature is — in his view would always elude physics.

[3]The momentum of a particle of mass M moving with the velocity v is given by $p = Mv$, unless the velocity approaches that of light, in which case the theory of relativity kicks in. You need not concern yourself about that.

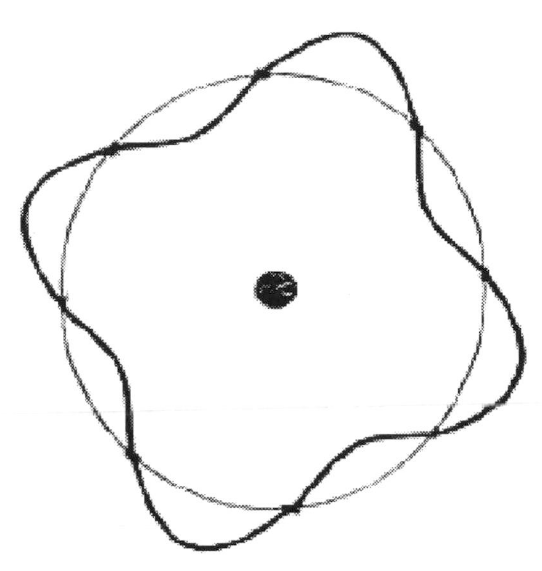

Figure 5.1. de Broglie's idea of Bohr's electronic orbit in atoms as standing waves.

by the Scottish physicist George Thomson. Performing essentially the same kind of experiment by which Thomas Young had proved the wave nature of light — though their experiment was much more difficult than Young's because the de Broglie wavelength of their electrons was very much shorter than that of light — they all found that beams of electrons exhibited quite similar interference effects showing that these 'particles' too had a wave-like nature. De Broglie's idea was not only correct, but it also fell on fertile ground.

The mid-nineteen twenties was the time when several young physicists made serious attempts at fleshing out Bohr's old ad hoc quantum theory of atoms and structuring it into a coherent theory. Stimulated by de Broglie's wave-notion for particles, Erwin Schrödinger sat down to devise a differential equation that would govern these waves, just as Maxwell had invented his differential equations to govern Faraday's light waves.

Born in 1887 in Vienna, Schrödinger studied theoretical physics at the University of Vienna, where he received his doctorate. During the First World War he served as an artillery officer in the Austrian army in a post where he had enough free time to read philosophy and physics — especially about Einstein's general theory of relativity. After the war and a number of temporary positions, he became Professor of Physics at the University of Zürich, where he remained for six years until he was offered, and accepted, the professorship vacated by the retirement of Max Planck in Berlin. It was in Zürich that he had his inspiration about the quantum theory.

The differential equation he postulated for the 'wave function' governing a particle came to be known as the Schrödinger equation. His resulting 'wave mechanics' competed with two apparently utterly different formal mathematical architectures for the quantum theory, one designed by the German physicists Werner Heisenberg and Max Born called 'matrix mechanics' and another by the British physicist Paul Dirac. Though all three theories were soon shown to be mathematically equivalent and called simply quantum mechanics, Schrödinger's remained the most popular among physicists because it employed familiar mathematical tools and lent itself most easily to actual calculations.

Applicable to particles of all kinds, free or subject to forces and confined, the Schrödinger equation of course had its simplest form for a free particle of mass M and momentum p, whose energy therefore consists of nothing but kinetic energy, which classically is given by $E = p^2/2M$. Its solution in that case is sinusoidal both in its behavior as a function of the time, with the frequency $f = E/h$, and as a function of distance, with the wavelength $\lambda = h/p$. It thus satisfied both Einstein's quantum requirement and de Broglie's wave postulate. Figure 5.2 shows the wave functions of the two states with the lowest energy of the bob of a simple pendulum.

There remained, however, the question: what was the meaning of Schrödinger's wave function? Schrödinger himself at first tended to think of it as the extended shape of a particle, replacing the idea of a point particle. This notion could not be maintained, because it turns out that, according to his equation, the wave function spreads quite rapidly all over space. The interpretation eventually adopted

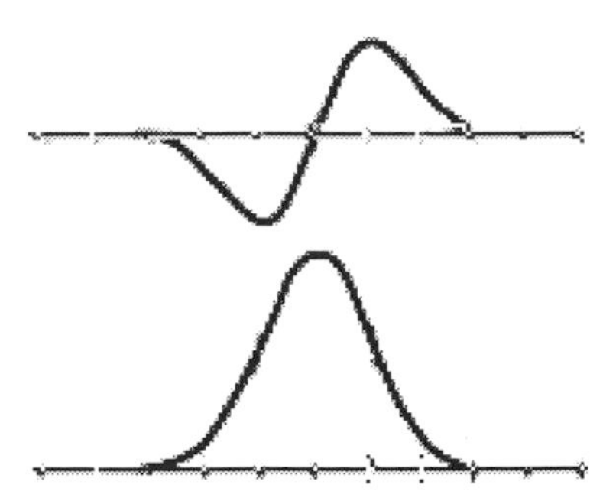

Figure 5.2. The wave functions of the two lowest states of a simple pendulum bob.

was proposed by Max Born: the square[4] of the value of a wave function in a given small region indicates the probability of finding the particle there. Quantum mechanics thus, much to the dismay of many physicists, became a fundamentally probabilistic theory.

Unhappy or not, most physicists simply ignored questions of interpretation, and found that the theory was phenomenally successful both qualitatively and quantitatively. For most of its numerical predictions, such as the precise energies of the electrons in atoms, the wave nature of the solutions of the Schrödinger equation and their intuitive interpretation was really irrelevant. But there were some important exceptions. Here is an example.

Certain heavy atoms are subject to a kind of radioactivity called alpha decay, emitting rays consisting of alpha particles. These are positively charged helium ions, made up of two protons — positively charged hydrogen ions — and two neutrons, all bound together by the strong nuclear force. As a result of their particularly strong binding, they stay together even inside the nucleus of a heavy atom, which consists of many protons and neutrons, all kept in a cage, so to speak, surrounded by a high barrier. This barrier is formed by the

[4]More accurately, since the wave function is a complex number, it is the square of its absolute value that is the probability. If you don't know what a complex number is, it really doesn't matter, but you may consult my book, *What Makes Nature Tick?* Harvard University Press, 1993.

strongly attractive but short-ranged nuclear force and the weaker but much longer-ranged repulsion between the positively charged alpha particles. The combination of these two forces produces an effective wall of a certain calculable height and width: inside the barrier the alpha particles are kept in a cage, but outside they would be repelled and fly away.

If they were just classical particles, there would be no escape for the alpha particles, as none has enough energy to surmount the wall, and the nucleus would be stable. However — just as sound waves are able to penetrate walls impervious to solid objects — as these are probability waves stretching to infinity, they have a certain calculable probability to be found outside: to tunnel through the wall. One can actually calculate the rate at which alpha particles in a given nucleus — and therefore having a given average energy there — will escape per unit time: the rate of alpha decay of that kind of nucleus. Such calculations were first done independently by the three physicists, George Gamow, Ronald Gurney, and Edward Condon, and they agreed with laboratory observations. Alpha radioactivity can thus be explained, both qualitatively and quantitatively, by wave mechanics.

Gravitational Waves

Just as Maxwell's equations imply that all accelerating electrically charged objects emit electromagnetic radiation in the form of waves, so Einstein predicted that his equations of the general theory of relativity — which is a theory of gravity — imply the existence of gavitational waves emitted by all massive accelerating objects. (By contrast, Newton's theory of gravity, an instantaneous, action-at-a-distance theory, did not. However, ever since Sir Arthur Stanley Eddington's verification in 1919 of the exact amount by which starlight was bent by the gravity of the sun, general relativity has superceded Newton's.) These waves should travel with the speed of light c, and just as in the case of light waves, their wavelength λ and frequency of oscillation f should be related by $f\lambda = c$.

When a gravitational wave strikes any object — since all objects have mass, they are subject to gravity — it produces a force on it; however, for all such waves that can be expected to be observed, the magnitude of this force would be extremely small and difficult to observe. That's because the only accelerating sources massive enough to produce sizable gravity radiation — for example, binary star systems rotating in orbits about one another — are extremely far away and the amplitudes of their waves therefore much attenuated. The method employed to detect them depends essentially on placing two very heavy solid cylinders a large distance apart, wired to detect the slightest vibrations by means of laser interferometry — interference effects in laser beams directed at them. (They are placed far apart so as to filter out vibrations locally caused by heavy traffic or Earth tremors.) The oscillation in synchrony of both the distant masses would signal the passing of a gravitational wave.

So far, no direct observational evidence for gravitational waves has been found, but large-scale LIGOs — an acronym for **L**aser **I**nterferometer **G**ravitaional-Wave **O**bservatory — have been set up by scientists at Caltech, MIT, and other universities. They are located about 3000 kilometers apart at the Hanford Nuclear Reservation in Washington State and in Livingston, Louisiana, waiting for the detection of coordinated vibrational signals. Other such projects are in the planning stage.

Although no direct evidence for gravitational waves has been detected so far, there is good indirect evidence for their existence. In 1974 the two American astronomers Russel Alan Hulse and Joseph Hooton Taylor working at the Arecibo Observatory in Puerto Rico discovered a binary system of two neutron stars, one of them a pulsar, orbiting each other. Careful measurements showed that their orbital period kept on gradually decreasing by about 76 millionths of a second per year, a fact that could only be explained by assuming they were losing energy by emitting gravitational radiation, as predicted by general relativity. (This is the precise analog of the electromagnetic radiation emitted, albeit at a much greater relative rate than the

Hulse–Taylor stars emitted gravity radiation, by electrons orbiting a nucleus according to Maxwell's equations. This fact would make the atom unstable if it were a classical system, with electrons orbiting the nucleus as Rutherford had proposed. That's what led Niels Bohr to propose his revolutionary quantum theory to stabilize it. There is no corresponding way to prevent those two neutron stars from eventually crashing into one another.) So there is little doubt that gravitational waves do exist, but it will be good to have direct evidence from LIGO.

Now to a more down-to-earth subject: water waves.

6

Water Waves

Waves in Shallow Water

The surface of a lake as well as of the ocean is normally kept flat by a force called surface tension, which is caused by the weak mutual attraction of neighboring water molecules. This surface tension produces a two-dimensional analogue of the restoring force that keeps a string tight when fixed at its ends. So when the water surface is disturbed, there are oscillations of the same kind as those of a plucked string. These ripples, called capillary waves, have short wavelengths of a few centimeters, and are generally of little importance.

The more significant waves, both on the surface of the ocean as well as of ponds and rivers, are the gravity waves (not to be confused with gravitational waves discussed earlier), the restoring force of which is gravity rather than surface tension. These have wavelengths ranging from meters to hundreds of meters. In shallow waters — meaning that the water depth D is much less than the wavelength λ — the velocity with which the waves travel is independent of λ and proportional to the square root of D. There are, however, some much more interesting kinds of waves in shallow water.

John Scott Russell was a Scottish naval engineer who taught mathematics at Edinburgh University but, in order to determine the best design for canal boats, he also performed a variety of experiments. In 1834 he made a remarkable observation which he described in the following words:

> I was observing the motion of a boat which was rapidly drawn along a narrow channel by a pair of horses, when the boat suddenly stopped — not so the mass of water in the channel which it had put in motion; it accumulated round the prow of the vessel in a state of violent agitation, then suddenly leaving it behind, rolled forward with great velocity, assuming the form of a large solitary elevation, a rounded, smooth and well-defined heap of water, which continued its course along the channel apparently without change of form or diminution of speed. I followed it on horseback, and overtook it still rolling on at a rate of some eight or nine miles an hour, preserving its original figure some thirty feet long and a foot to a foot and a half in height. Its height gradually diminished, and after a chase of one or two miles I lost it in the windings of the channel. Such, in the month of August 1834, was my first chance interview with that singular and beautiful phenomenon which I have called the Wave of Translation.

The unusual properties of these 'solitary waves,' or 'solitons' as they came to be called, are the following: Whereas normal waves would tend to either flatten out or steepen and topple over, these waves are stable, and can travel over very large distances. The speed of the wave depends on its size. Unlike normal waves they will never merge — so a small wave is overtaken by a large one, rather than the two combining. As these waves did not behave like any waves previously known, Scott Russell challenged 'future wave mathematicians' to find an explanation.

It took sixty years for such an explanation to be found. In 1894, by which time John Scott Russell was long dead, the young Gustav de Vries wrote his doctoral thesis at the University of Amsterdam under the supervision of Dieterik Johannes Korteweg, Professor of Mathematics. On the basis of the laws of hydrodynamics, he had derived what came to be known as the Korteweg–deVries

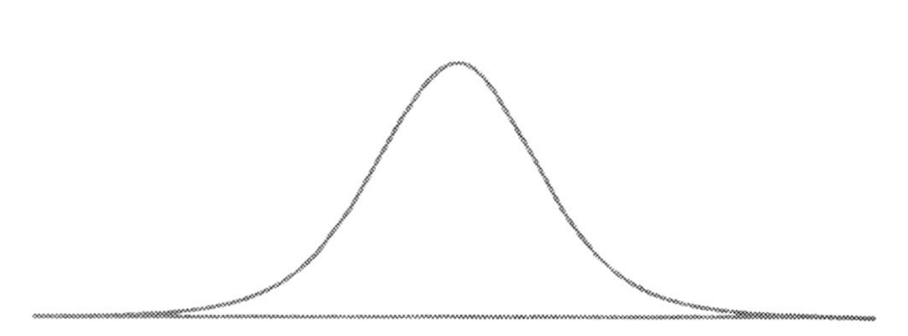

Figure 6.1. A solitary wave solution of the KdV equation.

(also simply KdV) equation that governed the water waves Scott Russell had observed. It is now known to apply also to waves in many other, very specialized contexts, such as waves in elastic rods, certain hydromagnetic waves, acoustic waves in anharmonic crystal lattices, ion-acoustic waves in cold plasmas, pressure waves in liquid–gas bubble mixtures, and others. Figure 6.1 shows the shape of a solitary wave solution of the KdV equation. The speed with which it travels is proportional to its amplitude.

In contrast to the well-known wave equation underlying sound waves as well as electromagnetic waves we have already discussed, the KdV equation is a nonlinear differential equation, which means that the superposition principle does not hold for its solutions, i.e., if f and g are two solutions, it is generally not true that their sum $f + g$ is a solution as well. It also contains no constant that determines the propagation speed of its wave solutions. For many years there was only one kind of solution known, the shape of which is shown in Fig. 6.1. (Nonlinear differential equations are much harder to solve than linear ones.) This changed after the development of the digital computer.

In 1965 a group of applied mathematicians under the leadership of Martin Kruskal at Princeton University performed a number of computer experiments to see what would happen to two solitary-wave solutions of unequal amplitude — and therefore of unequal speed — in the course of time. When they are far apart, so that each wave has its peak where the other is almost zero, their superposition very approximately solves the equation. However, if the larger one is behind, it eventually catches up with the smaller one, and the two begin to interact — in other words, the equation's solution is not simply the sum of the two original shapes but is more complicated. Remarkably enough, however, after a while a solitary wave of the same shape and amplitude as the original larger one emerges in front — slightly farther along than it would have been had the other not been there — with the smaller, slower one behind, slightly delayed. (See Fig. 6.2.) The two waves almost seem to behave like two particles with their own identity. The same thing occurs when any number of solitary waves of unequal speed collide, so to speak: they all eventually emerge unscathed, their identity preserved. This is what made solitons such fascinating phenomena.[1]

Are there equations analogous to the KdV equation in more than one dimension that have solitary-wave solutions like solitons? There

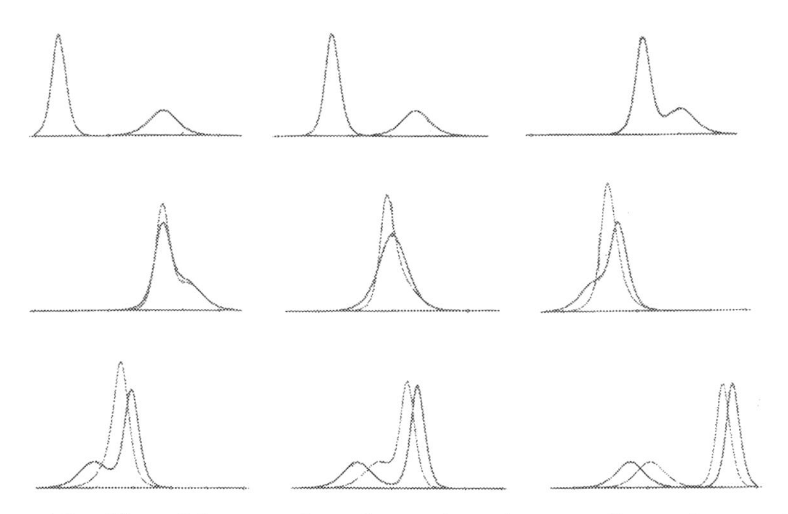

Figure 6.2. The solid curves show the motion of a two-soliton solution of the KdV equation; the dotted curves are the superpositions of the two corresponding solitary waves as they would move without the other one present.

[1]We shall not describe here the equally fascinating method found for systematically finding all the solutions of the KdV equation. For those who are interested, see my book *Thinking about Physics*. Princeton University Press, 2000.

has been some very limited success in finding such equations in two dimensions, but no success at all in higher dimensions. Remember also that these remarkable waves have been found and explained only in shallow water. What about waves in deep ocean water?

Waves in Deep Water

The biggest of the ocean waves are of course the tides. The primary cause of the tides is the gravitational attractions of the Moon and the Sun, which have the effect of tending to distort the shape of the Earth where it can, that is, where the Earth is plastic enough to allow noticeable distortion. (It was Isaac Newton who first explained this correctly, using his theory of gravity.) That, of course, is the ocean. So when the Moon or the Sun are situated over an ocean area they create enormous bulges that will tend to follow the Moon's and the Sun's motion. In the middle of the ocean this large and high wave is not really visible from a ship, because there is nothing to compare it with. Its wavelength is much too large for both the peak and the trough to be visible simultaneously. However, when its crest approaches a shore, the effect is very noticeable. Because the relative positions of the Sun and the Moon vary in the course of every month, so do the tidal periods as well as their amplitudes. When the Sun and the Moon are in line with the Earth, the amplitude of the tide is at its peak, called the *spring tide.* (This name has nothing to do with the season but originates from the word 'jump.') In between there are the *neap tides.*

The most commonly observed waves on the ocean, the ones that may cause even a large ocean liner to tilt and oscillate enough to make you sea sick and, in the past, to cause many smaller ships to capsize, are gravity waves of wavelength ranging from a few meters to hundreds of meters. Generated by winds and their turbulence, their shape is approximately sinusoidal unless their amplitude is large compared to their wavelength. Figure 6.3 shows the more peaked shape of the waves when their amplitude is larger, as well as their breaking form

Figure 6.3. The famous woodcut *The Great Wave off Kanagawa* by Hokusai (from Wikipedia).

Figure 6.4. Photograph of a tidal wave in the ocean.

for much larger amplitudes. During a severe storm, the wave height from trough to crest can be 30 meters or more.

These waves move with a velocity v that is proportional to the square root of their wavelength λ, and their frequency f of oscillation is given by $f = v/\lambda$. Just as in the case of sound, the motion of the waves does not mean that the water molecules move in the same direction. In fact, the water particles move in a vertical circular motion near the surface, their direction forward at the wave crest and backward at the trough; deeper down, they are mostly unaffected by the wind-caused surface waves.

Finally, there are the very large waves called tidal waves, caused either by hurricanes or by earthquakes on the ocean floor. The latter are usually called tsunami. So we shall finally return to our starting point, the tsunami, in the next, and last chapter.

7

And Back to Tsunami

The nature of tsunami is not entirely understood. One way to describe them is as ordinary water waves with lengths of hundreds of kilometers. These wavelengths are very large even when compared to the depth of the ocean. (The mean depth of the Pacific ocean is 4.3 km.) One may therefore be tempted to describe them like waves in shallow water, which would lead to a speed — proportional to the square root of the water depth — of about 740 km/h, or 462 miles/hour! In the open ocean such waves would have relatively small amplitudes and be hardly noticeable by passing ships. But as they approach a continental shelf, where the depth of the water becomes shallower, they would drastically slow down and therefore their amplitude would have to enormously increase to accommodate the mass of water they contain. The result at the shore would then have the destructive power of a tsunami.

While this description may well be appropriate for tidal waves, there is, however, also a quite different theory to explain tsunami. Rather than seeing them as ordinary, conventional water waves as in the above paragraph, they may be more akin to the 'heaps of water' or solitary waves that John Scott Russell had observed in his Dutch canals. The problem with this is that here we are dealing not with one-dimensional shallow canals but two-dimensional surfaces of deep oceans.

There is now ongoing research, both experimentally and theoretically, to understand the behavior of solitary waves in two dimensions, with a particular view to explaining tsunami by finding the appropiate mathematical equations to describe their behavior. (I am referring especially to the work of Professor Harry Yeh at Oregon State University and his collaborators.[1])

We have examined here the varieties of sound waves, the waves making up light and their sometimes spectacular effects, the waves underlying quantum mechanics, as well as the various kinds of normal water waves. In conclusion we have to say that tsunami are different from the usual waves we encounter everywhere, different not only because of their enormous destructive power but also because of their seemingly unusual nature. However, exactly what laws they follow and what their underlying equations are still remains to be discovered. In that sense tsunami still remains unexplained.

---

[1]Wewen Li *et al.*, *On the Mach reflection of a solitary wave: revisited* J. Fluid Mech. (2011), vol. 672, pp 326–357; Harry Yeh *et al. The 11 March 2011 East Japan Earthquake and Tsunami: Tsunami Effects on Coastal Infrastructure and Buildings*, Pure Appl. Geophys. 2012, Springer Basel AG.

References

Abbott, P. L., *Natural Disasters.* New York: McGraw-Hill, 2009.

Beranek, L., *Concert and Opera Halls: How They Sound.* Acoustical Society of America, 1996.

Heilbron, J. L., *The Dilemmas of an Upright Man: Max Planck as Spokesman for German Science.* University of California Press, 1986.

Karwoski, G. L., and J. MacDonald, *Tsunami: The True Story of an April Fool's Day Disaster.* Plain City, OH: Darby Creek, 2006.

Kuttruff, H., *Acoustics: An Introduction.* London and New York: Taylor and Francis, 2007.

Newton, R. G., *What Makes Nature Tick?* Cambridge: Harvard University Press, 1993.

Ohanian, H. C., *Physics.* New York and London: W. W. Norton & Co.. 1989.

Parker, B., *The Power of the Sea: Tsunamis, Storm Surges, Rogue Waves, and Our Quest to Predict Disasters.* New York: Palgrave-Macmillan, 2010.

Powers, D. M., *The Raging Sea: The Powerful Account of the Worst Tsunami in U.S. History.* New York: Citadel Press, 2005.

Thomas, M. E., *Optical Propagation in Linear Media: Atmospheric Gases and Particles, Solid-State Components, and Water.* Oxford University Press, 2006.

Thompson, J., *Cascadia's Fault: The Coming Earthquake and Tsunami that could Devastate North America.* Berkeley: Counterpoint, 2011.

Wewen Li *et al., On the Mach reflection of a solitary wave: revisited,* J. Fluid Mech. (2011), vol. 672, pp 326–357.

Yeh, Harry, *et al. The 11 March 2011 East Japan Earthquake and Tsunami: Tsunami Effects on Coastal Infrastructure and Buildings,* Pure Appl. Geophys. 2012, Springer Basel AG.

Part II

Particles

Introduction to Part II

The first essay of this book was concerned with an aspect of one of the most basic subjects of physics, the dynamics of the world, namely waves, waves that arise as natural phenomena or govern human communication. This second essay is going to deal with another basic subject, namely the structure of the world: the fact that the world is made up mostly of particles.

How and when did we find out about this fact? These particles are so tiny that nobody has ever actually seen them. What kind of particles are they and what are their properties? It will turn out that, whereas for millennia philosophers had speculated about the existence of atoms, after scientific evidence for the existence of atoms was actually found, they were not simply indestructible little entities as had been thought, but they have a non-trivial structure, including a nucleus at the center.

After describing what nuclear physics is all about, we turn to the discovery of other particles, some of whose existence had been predicted by new theories. After that, the search for new particles really took off, and the instruments required for this search became enormously large. What are accelerators for, and why do they have to be so huge and expensive?

The search for new fundamental particles became so successful that physicists had to devise methods to make sense of this confusing conglomeration. They even had to learn unfamiliar mathematical procedures that had been invented for completely different purposes but turned out to be indispensable. In the end, the new classification methods became just as useful for physicists as the invention of the periodic table of the elements was for chemists, and they led to many new insights. All of this will be described in this essay in a way that should be accessible to readers with no mathematical background knowledge.

8
Atoms

Early Atomism

The idea that all matter in the world is made up of tiny indivisible particles moving in a void arose in various cultures over time. As a philosophy called atomism, it may have arisen in India as early as the sixth century BCE, although whether it did so independently of Greece is a matter of controversy.

In Greece, the search for the nature of matter had begun in the sixth century BCE with Thales of Miletos, culminating some 180 years later with a mysterious philosopher named Leucippus, about whom very little is known, except that he had a disciple named Democritus, born in 460 BCE in Abdera in northern Greece who lived to be 90 years of age and left a large body of writings. Unfortunately little of his work survived undamaged.

The revolutionary aspect of the atomistic philosophy espoused by Leucippus and Democritus was their acceptance of the void in which their physically indestructible particles of matter moved. Democritus envisioned various kinds of atoms, living forever, moving randomly, like motes in a sunbeam on a windless day. Underlying all the variety of shapes, sounds, and colors of the perceived world, he saw nothing but a collection of silent, colorless atoms. It was the necessary acceptance of the void as their realm that made the notion of atoms unacceptable to Aristotle. While Plato had notions of different geometrical shapes for different atoms, he of course was always thinking of ideal entities, not objects in the real world.

Needless to say, none of the postulates of atomism was based on evidence; it was a philosophy.

These fundamental ideas of the atomic constitution of all matter were transmitted, along with much else of classical Greek thinking, to the later, generally much less sophisticated philosophical thinkers of early medieval Europe by the long and very influential didactic poem *De rerum natura*, written in the first century BCE by the Roman poet Lucretius. They remained undeveloped during the long period in Christian Europe now known as the Dark Age. The alchemists essentially accepted the atomic theory in some form — atoms were not necessarily regarded as indivisible — and it was particularly the case for the seventeenth century professor of medicine in Wittenberg, Daniel Sennert, who was an influential proponent of atomism. What stood in the way, hindering its general acceptance, was not the notion of the void in which atoms moved — Sennert was a strong Aristotelian — but the resistance to what was regarded as a mechanical philosophy. Surely not all the phenomena perceived by our senses, the sounds, the colors, the smells, the tastes could be explained on the basis of mechanical motions, and atomism was a view of the world that tolerated no other explanations.

It wasn't until the seventeenth century that Robert Boyle, a genuine scientist in the modern sense — in those days they were called natural philosophers — strongly espoused the mechanical philosophy along with atomism and the transformation of the mumbo-jumbo of alchemy (the legendary Doctor Faustus was an alchemist), which had been used primarily to perform mysterious tricks mostly in unsuccessful attempts to produce gold from lead, into the physical science of chemistry.

A member of the Experimental Philosophy Club at Oxford, a group that eventually was chartered by Charles II to form the Royal Society of London for Improving Natural Knowledge, now known simply as the Royal Society, Boyle was an extremely careful experimenter who regarded his experiments as tools to discover new facts rather than merely to convince skeptics. He had built in his laboratory an air pump that was able to produce by far the best vacuum ever achieved anywhere. With this he finally disproved Aristotle's doctrine that 'nature abhors a vacuum,' and he established an important gas law still named after him. However, he never managed to convince Isaac Newton of the great value of the new science of chemistry. Sir Isaac, though he believed in atoms, continued to prefer alchemy.

Until the end of the eighteenth century, the belief in atoms and the fights over whether all matter was made up of these tiny building blocks were based entirely on philosophical speculation and ideological convictions about mechanistic, materialistic causes in the world. The transition from atomism as a philosophy unsupported by evidence to a scientifically based knowledge had to wait until the early nineteenth century for the English chemist John Dalton. It should not be surprising that this fundamental step was taken by a chemist, as this relatively young science, more than any other, dealt with transformations of matter in the forms of solids, liquids, and gases.

Evidence for Atoms

A Quaker born in 1766 in Cumberland, the son of a weaver, John Dalton's early desires to study law or medicine were frustrated by being told that English law barred Dissenters — nonmembers of the Anglican Church such as Quakers — from attending or teaching at English universities. After moving to Manchester and acquiring much of his scientific knowledge through informal instructions from a blind philosopher and polymath, John Gough, he was appointed teacher of mathematics and natural philosophy at the dissenting "New College" in Manchester. He then resigned his post and became a private tutor in the same subjects with the addition of chemistry.

While Dalton's main interest started out to be meteorology, he made contributions to many different areas, including color blindness (with which he was afflicted himself). However, his most important work was done when the focus of his interest changed from meteorology to chemistry after becoming convinced that all substances were made up of particles. It was the observation that liquids such as water were able to absorb gases, important in meteorology, which persuaded him that this must be so. Water could absorb air only, he thought, because the gas particles were able to occupy the interstices between the water particles. If there were no particles, there would be no empty spaces enabling absorption.

Chemistry presented Dalton with the crucial chemical fact, already well recognized by Robert Boyle, that, in contrast to mixtures, chemical compounds — substances that generally had quite different chemical properties than their individual constituents — were made up of elements in very specific ratios of weight. This was known as the law of definite proportions. Dalton added what became known as the law of multiple proportions: if two different elements can combine with one another in more than one compound, their weight ratios in the various cases always differ by factors that are small whole numbers. Here are two examples.

The elements carbon and oxygen can combine in two different ways. The first compound (now called carbon monoxide) contains 42.9% by mass carbon and 57.1% by mass oxygen; the second compound (now called carbon dioxide) contains 27.3% carbon and 72.7% oxygen. The mass ratio of oxygen to carbon in the first is $57.1 : 42.9 = 1.33$ and in the second $72.7 : 27.3 = 2.66$. So the ratio of the two is $2.66 : 1.33 = 2$.

The elements nitrogen and oxygen are able to combine in five different ways to form compounds; in these five oxides, 14 grams of nitrogen combine with 8, 16, 24, 32, and 40 grams of oxygen respectively, which differ by factors of 2, 3, 4, and 5.

This law confirmed Dalton's conviction that each element must be made up of particles — atoms, though they need not be indestructible — and, furthermore, that these particles are all alike and have the same weight, which in chemical compounds attach themselves to one another to form what are now called molecules (a name invented by the French chemist Joseph Louis Gay-Lussac). Furthermore, he concluded from the law of definite proportions that if two elements were able to form a compound with a given weight ratio, this ratio must be equal to the ratio of the weights of their respective atoms. When molecules contained more than one atom of the same element this reasoning was of course wrong, and he was still a bit confused about the difference between atoms and molecules when it came to some gases made up of single elements, such as oxygen, O_2, the molecules of which consist of two identical atoms. He was wrong about some atomic weights[1] and the constitution of some molecules, but he certainly grasped the important main idea. Finally, atomism had advanced from philosophical speculation to a theory, not always correct in all details but now with a scientific basis.

John Dalton, always an outsider who never married, was eventually elected to the French Academy of Sciences and — over his objections — to the Royal Society, receiving the Royal Medal, and honorary degrees from the Universities of Oxford and Edinburgh. After retiring with a government pension, he died in Manchester in 1844.

It did not take very long for Dalton's ideas to be refined by the Italian physicist Avogadro.

Amedeo Carlo Avogadro was born in 1776 in Turin in Piedmont, Italy, to a noble family (as an adult he was Count of Quaregna and Cerreto). After graduating with a degree in ecclesiastical law, he began a practice, but changed his mind and dedicated himself to *positive philosophy*, which meant physics and mathematics. Soon after beginning to teach, at the age of 32, at the Royal College of Vercelli, he published an article that made his reputation,[2] and in 1820 he was appointed the first professor of mathematical physics at

[1] What we now call the atomic weight of an element is the ratio of the weight of one of its atoms to that of an atom of the lightest element, hydrogen.

[2] It was written in French and published in a French journal. At that time France controlled northern Italy.

the University of Turin. However, when he became politically active in the revolutionary movement against King Victor Emmanuel I — who had come to power after the downfall of Napoleon when France lost control of northern Italy — he was removed from his professorial chair for ten years. After returning, he taught for another twenty years. Later regarded as the founder of physical chemistry, he was a very modest man who never, during his lifetime, received the recognition he deserved for his important work. He died in Turin in 1856 at the age of 79.

Avogadro's most significant achievement was the formulation of a rule that Dalton had already considered but rejected because its consequences violated his intuition that the heavier molecules in a gas were bigger than the lighter ones and that they were all tightly packed together with small interstices. But Avogadro came to a different view based primarily on what is known as Charles's law[3]: for a given rise in temperature, all gases rise by the same fraction of their volume. Avogadro's conclusion from this is contained in his important law (which for many years, sometimes even today, is referred to as Avogadro's hypothesis): *at any given temperature, all gases contain the same number of molecules per unit volume.* It allowed him to explain rationally why Gay-Lussac's law was true: when two gases combine chemically, their volumes and those of their product (if it is a gas) at a fixed pressure always stand in simple numerical ratios. (For example, two gallons of hydrogen and one gallon of oxygen yield two gallons of steam.) It was clear that Dalton's intuition evidently had been wrong; the spaces between the molecules of a gas had to be relatively large.

Perhaps the most important consequence of Avogadro's law was that it could be used to find the relative weight — relative to that of hydrogen — of each individual molecule of a gas directly from

[3]Named after the French physicist Jacques Alexandre César Charles, it was independently discovered five years later — Charles had not published his discovery — by Gay-Lussac.

the density of that gas. This ratio must be equal to the ratio of the weight of a unit volume of that gas to the weight of a unit volume of hydrogen, that is, the ratio of the density of the former to the density of the latter, a method much more reliable than the one Dalton had used.

By defining a *gram mole* as a quantity of gas whose weight in grams equals its relative molecular weight, Avogadro's law can also be stated in terms of weights rather than volumes (with no need to mention temperature): the number of molecules in a gram mole of any gas is a universal constant, which is now called Avogadro's constant. For example, this is the number of molecules in 32 grams of oxygen as well as in 2 grams of hydrogen, as well as in 18 grams of water (remember water is H_2O), since each molecule of hydrogen gas contains two hydrogen atoms, each molecule of oxygen contains two oxygen atoms, and each molecule of water (steam) contains two hydrogen atoms and one oxygen atom. Thus two gallons of hydrogen and one gallon of oxygen make two gallons of steam. However, Avogadro did not give the numerical value of his constant, which remains one of the most important fundamental constants of chemistry and physics. The determination of this number, 6.022141×10^{23}, had to wait until the end of the nineteenth century, when more was known about the constitution of molecules.

After the work of Dalton and Avogadro, chemists, of course, had to accept that all the material they handled consisted of atoms and molecules. It transformed their whole laboratory practice from nothing but smelly and dangerous work with bunsen burners and test tubes to one more akin to that of physicists. For physicists, the path toward the acceptance of atoms was the understanding of the nature of heat.

The development of the science of thermodynamics was one of the major scientific achievements of the nineteenth century. For millennia, heat — the ancient Greeks called it fire, one of the four elements, along with earth, water, and air — was considered a fluid that had the intrinsic property of flowing from a hot body to a colder one. But after it was realized that mechanical friction produced heat — the drilling of gun barrels made them red hot — and that heat melted ice, the fluid theory had to be abandoned. Then the recognition of the molecular constitution of gases and all material substances, as developed by Dalton and Avogadro, led to the kinetic theory, which postulated that heat was nothing but a constant irregular motion of the constituent molecules. The faster that motion, the hotter the substance: the heat energy of a gas was equal to the average kinetic energy of the particles making up the gas. Thus atoms and molecules became basic concepts in physics just as in chemistry. This was the context in which the physicist and chemist Johann Josef Loschmidt made his important contribution.

Josef Loschmidt, as he called himself, was born in 1821 to a peasant family in Carlsbad (now Karlovy Vary, the Czech Republic) in the Austrian Empire. After a priest persuaded his parents to send him to high schools at a monastery and then in Prague, he studied philosophy and mathematics at the Charles University in Prague. His best known paper, *Zur Größe der Luftmoleküle* (On the magnitude of air molecules), gave a remarkably accurate estimate of the size of the molecules that make up the air. Using all the tools of the then well-developed kinetic theory of gasses, he calculated the number of molecules per cubic meter at standard temperature and pressure[4] that differed only by a factor of two from the now accepted value of 2.686763×10^{25}. After many other path-breaking contibutions to chemistry, Loschmidt retired in 1891 and died in Vienna in 1895.

So one of the important effects of the kinetic theory of gasses was to advance the idea of atoms and molecules closer to reality, as it was, after all, one of the primary tools of that theory. But the strange notion of entropy presented some difficulties. Here was a concept that, from a mechanical point of view, was very hard to understand. The first law of thermodynamics — conservation of energy — had cleared up what earlier had been quite puzzling, namely that in the

[4]This number is now called Loschmidt's constant.

presence of friction mechanical energy was not conserved. Now that heat was understood to be a form of energy, this puzzle was solved. The second law of thermodynamics, which states that the entropy of a closed system could never decrease — it could only remain constant or increase — was very hard to reconcile with the idea that physical systems consisted of particles such as atoms and molecules, all of which obeyed the Newtonian laws of motion. These laws make no distinction between the directions of time. If you took a video (forget that this is a bit anachronistic) of the motion of a planet and ran it backwards, there is no way of telling that it was running the wrong way; what it shows is equally possible. There are, however, physical processes that are irreversible, a video of which run backwards would look ridiculous; for example, a raw egg dropped from the edge of a table and landing on the kitchen floor, or a gradually cooling cup of coffee. Thermodynamics deals with these phenomena by means of entropy, but how can particles account for them? The man who managed to solve this deep and challenging puzzle was a young student of Loschmidt's in Vienna.

The Reality of Atoms

Born in 1844 in Vienna, his father a government official, Ludwig Boltzmann attended high school in Linz, after being privately tutored, and studied physics at the University of Vienna, obtaining his PhD in 1866. Three years later he was appointed Professor of Mathematical Physics at the University of Graz and in 1873 he accepted the position of Professor of Mathematics at the University of Vienna, but returned to Graz to accept the Chair of Experimental Physics, where he did his most important work. After moving, six years later, to the University of Munich, he returned to the University of Vienna in 1893 as Professor of Theoretical Physics, but he also became a popular lecturer on philosophy. His entire life Boltzmann was subject to periods of severe depression and rapid changes of mood, probably undiagnosed bipolar disorder. During a summer vacation at Duino near Trieste in 1906, his depression overwhelmed him and he hanged himself.[5]

Founding the new science called statistical mechanics, Boltzmann's basic insight was that since the number of molecules in a laboratory container of gas was so enormously large, there was no way — or interest, for that matter — to follow their individual motions. The observable, measurable properties of the gas, such as its temperature and pressure, depended only on statistical averages of its huge number of constituents. In other words, the laws governing gases and fluids had to be explained in terms of probabilities and statistics applied to a vast number of tiny interacting particles none of which could be individually identified. The famous formula he promulgated, stating that the entropy of the state of a system is proportional to the logarithm of its probability, $S = k \log W$, is engraved on his tombstone. (W stands for the German word 'Wahrscheinlichkeit' for 'probability'.)

For example, when you open the door between two rooms, in one of which the air is quite hot — the molecules, on average, are moving very rapidly — and in the other it is much cooler — so its average molecular motion is much slower — then the probability is overwhelmingly in favor of the two large bodies of molecules becoming mixed so that those in the hot room, on average, move more slowly than before and those in the cool room, on average, more rapidly: the hot room cools down and the cool room heats up. Although in an extremely small number of instances it may happen that the hot room heats up — by sucking up, so to speak, the faster moving molecules of the cool room — and the cool room cools down — many of the slower molecules in the hot room moving there — the probability for such an occurrence is vanishingly small. Thus irreversibility is explained by means of the fact that, although matter is made up of atoms and molecules, each of which follows

[5]It happened to be a place made famous a few years later by the Austrian poet Rainer Maria Rilke in his *Duino Elegies*.

Newtonian laws, these constituents are so tiny and their number is so enormous, that the observable behavior of matter is based on the laws of statistics. This was Boltzmann's deep insight.

But were these particles real? Boltzmann's final goal, to which he dedicated his entire life, was to convince the world that atoms were real. And there he had a big fight on his hands, particularly in his homeland. His main enemy — and these attacks did become very heated and personal — was his colleague, the physicist Ernst Mach, Professor of the History and Philosophy of Science at the University of Vienna, who espoused and propagated the philosophy of positivism.[6] Mach did not deny the validity of the kinetic theory of gases, but he insisted that just because atoms and molecules were convenient postulates for the theories of physicists and chemists, that was not a sufficient reason to believe in their reality. Who had ever seen an atom? There is no reason to believe that Boltzmann's suicide was caused by his unhappiness on account of these relentless attacks, but they surely added to his misery. Sad to say, the year before his suicide, Einstein had published a paper on Brownian motion, the experimental verification of which definitely established that atoms are real, but Boltzmann never knew about it.

Here is the story. The man's name was Robert Brown, a Scottish botanist, born in 1773 in Montrose, the son of a minister in the Scottish Episcopal Church, and educated at the University of Edinburgh. He did research largely by means of his microscope.

In 1827, Brown was examining pollen grains suspended in water, and he noticed that minute particles — starch and lipid organelles — ejected from them were constantly moving around in some sort of jittery dance. At first he thought the tiny objects must be alive, but then he found that the Dutch biologist Jan Ingenhousz some forty years earlier had already published that he had observed similar motions of small charcoal particles suspended in water, so life was

---

[6]He is now mostly remembered because the speed of an object travelling in a medium relative to the speed of sound in that medium is called its Mach number.

not the correct explanation. What became known as *Brownian motion* remained mysterious for more than three quarters of a century.

One of the four path-breaking papers Albert Einstein wrote in 1905 dealt with Brownian motion. The cause of these mysterious movements of Brown's particles, he thought, was that they were continually bombarded by the irregular motions of the water molecules around them. Now, this idea by itself was not new, but Einstein took it seriously and calculated in detail the statistical properties of the random motions of microscopic particles, such as Brown had observed, resulting from their collision with the still controversial sub-microscopic water molecules of a known mass and at a given temperature as given by thermodynamics. Three years later, these random motions calculated by Einstein were verified under his newly developed ultramicroscope by the French physicist Jean-Baptiste Perrin, thereby definitely demonstrating the reality of water molecules and clinching the case for the existence of atoms. On the basis of these observations and the kinetic theory of gases, Perrin was also able to estimate reliably the size of atoms and molecules. In 1926, he won the Nobel Prize in physics.

What are Atoms like?

By this time the original idea that atoms were indivisible entities with no structure had long been dropped and the question naturally arose, what were they made of, what were they like? They could of course not be seen under any microscope, as they were much too small. So reasonable models had to be invented. For that we have to backtrack a little.

In 1897, the German physicist Ferdinand Braun had invented what was became known as the Braun tube but later as a cathode ray tube or CRT. It is a glass tube with electrical connections in the back and a wide phosphor-coated screen in front. The positive terminal, called the anode, is separated from the negative one, called the cathode, and the latter is heated[7] like the filament in a light bulb and was found to emit certain rays that become visible when they strike the front screen. But what were these rays? It was the English physicist J. J. Thomson whose experiments demystified the cathode rays.

While many thought cathode rays were immaterial disturbances in the ether, as light was thought to be, others believed them to be made up of particles. The experiments J. J. Thomson performed in 1897 showed definitely that they were made up of negative electrically charged particles. He proved this by finding that they could be deflected from their normal path by means of magnets as well as by means of external electric fields. In fact, the amount of deflection by means of a magnetic field of a given strength compared to the degree of deflection by given electric field allowed him (employing Maxwell's equations of electromagnetism) to calculate the ratio of their electric charge to their mass. When compared to the (positive) charge-to-mass ratio of a hydrogen ion, it was more than a thousand times higher. Since at the time there was already reason to believe that the charge of this 'corpuscle,' as Thomson called it, was equal and opposite to that of the hydrogen ion, the mass of the electron, as it was later called — and still is — was less than one thousandth of that of the hydrogen ion. (The ratio is now known to be about 1/1836.)

Joseph John Thomson was born in 1856 in Manchester, England. He was educated in a small private school and at the young age of 14 years admitted to Owens College. After his father died in 1873, he moved on to Trinity College, Cambridge, where he obtained his B.A. in mathematics in 1880 and his M.A. in 1883. A year later he became Cavendish Professor of Physics. He was a highly gifted teacher and made numerous important contributions to physics, he was elected a Fellow of the Royal Society in 1884, became its President in 1915, won

---

[7] In Braun's original invention the cathode was cold, the heating was a later improvement.

the Nobel Prize in 1906, was knighted in 1908, and became Master of Trinity College in 1918. J. J. Thomson died in 1940 in Cambridge and was burried in Westminster Abbey.

The reason why the charge of the electron was believed to be equal and opposite to that of the hydrogen ion was based on the discovery of radioactivity in 1896 by the French physicist Enri Becquerel, and one of the rays emitted by radioactive elements was found to be cathode rays, subsequently named beta rays by Ernest Rutherford. If cathode rays consisted of electrons, these particles had to come from the inside of atoms. Hydrogen, the lightest element, had only one kind of positvely charged ion, so the electrically neutral hydrogen atom presumably contained just one electron, the loss of which made it into a hydrogen ion — ergo the charge of the electron had to be equal and opposite to that of the ion.

So J. J. Thomson devised a model of what he thought an atom looked like on the inside. He imagined it to be filled with some kind of positively charged soft heavy matter that made up almost the entire weight of the atom — the tiny electrons contributed almost nothing to the weight — and inside this squishy stuff there were the negatively charged electrons, so that the atom as a whole was electrically neutral. The model was generally called the plum-pudding model, as that's what it seemed to resemble. This is were matters stood when Ernest Rutherford entered the scene, coming from down under, the other side of the world.

Discovery of the Atomic Nucleus

Born in rural Nelson, New Zealand, in 1871, Ernest Rutherford was the second son of a total of twelve children of a Scottish father, a wheelwright and engineer, and an English mother who was a school teacher. After reading his first science book in elementary school, he managed to win a scholarship to attend secondary school at Nelson College, and at age eighteen he won one of ten scholarships nationally available to enroll at the University of New Zealand. After earning his B.A., he won the only Senior Scholarship in Mathematics available to continue on for the M.A. degree with First Class Honors in Mathematics, Mathematical Physics, and Physical Science. In 1894 he returned to Canterbury College to get the B.Sc. degree in order to be eligible for a new scholarship available to one candidate every other year from New Zealand to go anywhere in world to do research. Winning it, he chose to travel to the Cavendish Laboratory at Cambridge University to work with J. J. Thomson, becoming the first research student there who had not graduated from Cambridge University. He never returned to New Zealand, except once in 1900 to marry the woman he loved.

After successfully pursuing a number of other research directions, Rutherford decided to concentrate on understanding the newly discovered radioactivity of atoms. In 1898 he observed that these atoms emitted two quite distinct kinds of rays, which he named alpha and beta rays, the latter then found to be high-speed electrons, as mentioned above. Soon thereafter he accepted a professorship at McGill University in Montreal, Canada, where the research laboratories were very well equipped. In 1907, however, Rutherford was enticed to return to England and accepted a chair in physics at Manchester University. There he decided to study the atom, building a world-famous laboratory, second only to the Cavendish.

Together with a young chemist Frederick Soddy, he solved the mystery of how radioactivity appeared to create energy out of nothing, of how an element, after sitting inert for a very long time, could suddenly emit energetic rays. They found that each radioactive element had its own characteristic 'half-life,' after which half of the atoms in a given sample had decayed by emitting alpha or beta rays and at the same time turning into atoms of another element — it was a new form of resurrected alchemy — which may itself be again radioactive until the end product was a stable element, which in the case of radium was lead. When the half life of an element was extremely long — in some cases thousands of years — it appeared to be quite inert until it suddenly produced energetic rays. For this

Figure 8.1. The plum-pudding model of an atom envisaged by J. J. Thomson. The negatively charged particles are electrons.

discovery of 'transmutation' of elements he was awarded the Nobel Prize in chemistry in 1908.[8]

The atom-model that Rutherford meant to explore and test experimentally was of course the one proposed by J. J. Thomson, shown schematically in Figure 8.1. In order to do that, he told his two assistants, Hans Geiger and Ernest Marsden, to direct a stream of the positively charged helium ions emitted by a sample of the radioactive gas radon at a thin gold foil. (Rutherford called the helium ions alpha rays or alpha particles in contrast to the electrons emitted by some other radioactive substances, which he called beta rays or beta particles.) In order to find out how the electron plums were distributed in the pudding, they should carefully detect how many of these alpha particles were deflected in various directions. The result they reported was an enormous surprise: Some of them had been deflected in the backward direction! How could the tiny electrons deflect backwards alpha particles that weighed more than 7000 times as much? The mushy positively charged pudding of the

---

[8]It was a great surprise to him, as well as ironic, that his prize was in chemistry because he sneered at all sciences other than physics as 'no better than stamp collecting.'

atom surely couldn't do so either. J. J. Thomson's model had to be wrong; the positive charge of the atom had to be concentrated in a small heavy central nucleus, with the negatively charged electrons orbiting around it like the planets about the Sun.

In fact, the details of the way the beam of alpha particles was scattered in all the various directions by the repulsive force, called the Coulomb force, between them and the positively charged nucleus — called the scattering cross section — could be calculated by means of the Newtonian equations of motion, and the result agreed with what the Geiger–Marsden experiment showed.[9] The scattering observed even allowed Rutherford to calculate the size of this central nucleus that did the scattering. Its diameter could not be larger than about one 100,000th of that of the atom! "Like a fly in a cathedral" is the way Rutherford put it. Thus the atom, far from being a solid indestructible block, was almost entirely empty space.

So now he had an apparently very satisfactory model of the atom similar to the solar system. There was, however, a fatal flaw in this picture. According to Maxwell's equations of electrodynamics, each of the orbiting electrons would constantly emit electromagnetic radiation and lose energy. Thus they would all quickly spiral inwards and be swallowed up by the nucleus. Rutherford's atom was unstable; it could not exist for more than a fraction of a second.

In 1912 Rutherford hired a young Danish assistant named Niels Bohr, a man who had not only a lively imagination but who was a born revolutionary, ready to ignore universally accepted scientific doctrines and replace them with ad-hoc assumptions to be verified by new experiments. Bohr postulated a model of the atom resembling Rutherford's but restricting the electrons to specific designated orbits. Simply ignoring Maxwell's equations, he postulated that when an

---

[9]Remarkably enough the result calculated by means of classical mechanics happens in this exceptional case to coincide with the correct result obtained by means of the later developed quantum mechanics. Who knows how much Rutherford's discovery of the nucleus would have been delayed had this not been the case.

electron moved along one of these special orbits, it would not radiate. (See Chapter 5 for further remarks about these special Bohr orbits.) In each of these orbits the electron had a fixed energy, and unless it was in the orbit of lowest energy, called the ground state, it would, after a while, perform a 'quantum leap' and descend to one of lower energy, emitting the difference E in energies in the form of electromagnetic radiation. The frequency ν of this radiation emitted would be related to that energy difference by the formula that Max Planck had devised, $E = h\nu$, where is h is Planck's constant. This model accounted for the discrete spectrum characterizing the element. It was the beginning of the revolutionary quantum theory of atoms that transformed not only physics but chemistry as well. If the number of orbiting electrons in its atoms was assumed equal to its number in the periodic table of the elements, the model would eventually account for the properties of this periodic table which had been devised totally ad hoc by Mendeleev. This implied that the chemical properties of each element was determined by the number of electrons in its atoms. We shall return to the quantum theory anon.[10]

During the Great War, Rutherford worked for the British navy on methods of locating submarines, and after it ended he moved to Cambridge as professor of physics and director of the Cavendish Laboratory, succeeding J. J. Thomson. He was appointed professor of natural philosophy at the Royal Institution as well. In 1914 he was knighted and in 1921 elevated to the peerage with the title Lord Rutherford of Nelson. He served as president of the Royal Society from 1925 to 1930 and during the 1930s, as president of the Academic Assistance Council, he helped refugee scientists who had escaped from Nazi Germany. Regarded by many as the greatest experimental physicist of the twentieth century, Rutherford died in 1937 in Cambridge as a result of an accidental fall while gardening and was buried in Westminster Abbey.

---

[10]For a biography of Niels Bohr, see A. Pais, *Niels Bohr's Times in Physics, Philosophy, and Polity*. Oxford: Clarendon Press, 1991.

9

Nuclear Physics

Now that there was, finally, what appeared to be a satisfactory model of the atom, the next question of course was what the nucleus was like. In an atom of any given element, the number of electrons orbiting — more or less in the manner postulated by Niels Bohr — was determined by the positive electric charge of this tiny but heavy object at the center. What is more, in an atom of a radioactive element, the nucleus was evidently the source of this radioactivity. When Bohr's ad-hoc prescription was in the 1920s replaced by the architecture of a coherent theory that came to be called quantum mechanics, with his postulates now theoretically justified and found fundamentally correct (though not in every detail), this model became the basis of all of chemistry, as already mentioned at the end of the last chapter. But radioactivity with its regular emission of either alpha particles or electrons, and the later discovered fission and fusion, showed that this nucleus had to have some kind of structure, the exploration of which became a very active and fertile branch of physics for many years.

What did the atomic nucleus consist of? Hydrogen being the lightest element, and the atomic weights of all other elements being very approximately integral multiples of the weight of the hydrogen ion,[1] this ion was assumed to be simply a basic particle and was called

[1]The reason why they were not exact multiples was discovered later and will be discussed further on.

'proton.' There was only one ion of hydrogen, and its electric charge was equal and opposite to that of the electron, so its atoms could be assumed to consist of a proton as their nucleus and a single orbiting electron and the atom as a whole would be electrically neutral. What about the nuclei of the atoms of all the other elements? They could not simply consist of protons because the atomic weights of all elements other than hydrogen were larger — usually approximately by a factor of two — than their number in the table, so their nuclei had to be much heavier than the weight of the number of protons needed to equal the number of electrons. There was an easy way out: the nucleus simply had to contain also an appropriate number of electrons to neutralize the surplus protons. This was a band-aid solution of the problem and it lasted only until the discovery of the neutron in 1932 by the English physicist James Chadwick.

Chadwick, born in 1891 in Collington, Cheshire, had graduated from Manchester University, where he had studied under Rutherford. Because he had been pursuing research in Germany in 1914, he had been interned in a camp near Berlin during the Great War, where he was allowed to set up a scientific laboratory, and did not return to Manchester until 1918, moving to the Cavendish when Rutherford did and received his PhD in 1921. While doing research there in 1932, he learned that Walter Bothe and Herbert Becker in Germany, bombarding beryllium with polonium, had produced an unusual kind of electrically neutral radiation the nature of which they could not understand. Chadwick immediately recognized that this radiation would have to consist of heavy neutral particles the existence of which Rutherford and he had hypothesized for years: neutrons. As proof, he repeated the Bothe–Becker experiment, directing the mysterious radiation at parafin wax, where it displaced protons, which would then be detected by means of an ionization chamber and an oscilloscope (particles without electric charge are difficult to detect directly); the radiation had to consist of neutral particles heavy enough to be able to knock out protons. He had indeed discovered the neutron, the mass of which he and Maurice Goldhaber, a refugee physicist from Nazi Germany, subsequently determined experimentally to be almost equal to that of the proton, just slightly larger.

It was now clear: the nucleus of the atom of a given element consisted of a number of protons equal to the number of electrons orbiting around and thus equal to the element's number in the periodic table, and a number of neutrons to make up the element's atomic weight. (Protons and neutrons were from then on both referred to as 'nucleons.') Why then were the experimentally determined atomic weights not exact integral multiples of that of hydrogen? (The difference between the weights of the proton and the neutron was too small to account for this.)

It was found that all elements had several different *isotopes*, whose atomic nuclei contained the same number of protons but different numbers of neutrons. For example, 'deuterium' is an isotope of hydrogen whose nucleus, called a 'deuteron,' consists of a proton and a neutron. Normal water contains only a small percentage of molecules in which deuterium atoms take the place of hydrogen atoms. (However, the rare liquid called 'heavy water' contains a very much higher fraction of such molecules.) The chemical properties of two isotopes of the same element being identical, they could not be distinguished by chemical means, and those found in nature — those were the ones whose atomic weight was measured by chemists in the laboratory — were usually mixtures of several isotopes, so that the atomic weight obtained was usually not an integral multiple of the weight of the hydrogen atom.

Another neutral particle made its appearance in the laboratory, albeit indirectly. When an atom undergoes radioactive alpha decay, one can easily calculate the energy of the emitted alpha particles by means of the conservation laws of energy and momentum from the masses of the parent nucleus, the daughter nucleus, and the alpha particle. The same should be the case for beta decay, it was thought. However, in 1930, it was found that the electrons emitted by a given beta-radioactive element did not all have the same energy; their energies varied over a considerable range. This presented a real puzzle; Niels Bohr was even tempted to sacrifice the law of conservation of energy. There was, however, a young Austrian physicist named Wolfgang Pauli, who made the brilliant suggestion that the explanation was that in beta decay, the atom emitted not just an electron, but also another particle that remained undetected because it was electrically neutral. In that case the energies of the individual emitted particles were not fixed, only the sum of the energies of the two was. From the upper limit of the energy range of the emitted electrons — the energy of the emitted electron is near its maximum when the other particle is almost at rest and its energy is very close to its rest energy mc^2 — it could be deduced that this fugitive neutral particle had to be extremely light, much lighter even than the electron. The Italian physicist Enrico Fermi therefore named it 'neutrino,' the little neutral one.

Since this particle interacts with other matter only extremely feebly — almost all those emitted by the Sun and striking the surface of the Earth go straight through the entire globe — it is very hard to detect and after a long time, it was finally directly found in 1956 by the American physicists Clyde Cowan and Frederick Reines. By that time it was possible to create a very dense flux of neutrinos from beta decay, which was directed at a large tank of water, where some of them collided with protons and, by an inverse beta-decay, produced neutrons and positrons (a positively charged light particle, the discovery of which we shall discuss later). These positrons subsequently were annihilated by collisions with electrons, producing gamma rays, which were detected by photomultiplier tubes. They confirmed their finding by a second experiment using another method. Although by that time no physicist seriously doubted that Pauli was right and neutrinos did exist, it was reassuring to have it experimentally confirmed. We now know that there is more than one kind of neutrino. There is not only the kind of neutrino produced in beta decay in association with electrons, but in 1962 a different kind of neutrino, produced in asociation with another, heavier negative particle called the 'muon' which will be discussed shortly. So it was called the 'muon neutrino.' Not only that, but after the discovery of the tau particle in 1978 (also to be discussed later), it became clear that there has to be a third kind of neutrino, a 'tau neutrino,' which, however, has not yet been directly observed. All the neutrino masses have to be extremely small, though not zero. What is more, they can slowly be converted into one another; this is called 'neutrino oscillations.'

The biggest puzzle confronting nuclear physicists, of course, was what held the particles in the nucleus together. After all, the positively charged protons repelled each other, and the neutrons were neutral so there was no electrostatic force on them at all. There had to be an unknown kind of attractive force that provided the nuclear glue between the protons and between the protons and the neutrons. The answer was suggested by the Japanese theoretical physicist Hideki Yukawa.

Born in Tokyo in 1907, Yukawa received his degrees at Kyoto Imperial University, where his father was professor of geology. Interested primarily in the physics of elementary particles, he became assistant professor of physics at Osaka University, following with great interest the latest publications of the fundamental research in Europe. The hottest theory in physics in the 1930s, subject to intensive development in Europe at the time was quantum electrodynamics, also called QED, a quantum field theory destined to replace Maxwell's classical theory in the submicroscopic domain. In QED the quanta of the classical Coulomb force were photons, and because this force varied as the inverse square of the distance and therefore had an infinite range, the mass of the photon was zero. When Yukawa attempted to construct an analogous quantum field theory for the nuclear force, he realized that, although it had to be quite strong at short distances to overcome the Coulomb repulsion between protons, its range had to be more or less equal to size of the nucleus, because outside the nucleus this force had never been detected. According to the rules of quantum field theory, the quanta of this force, particles analogous to photons — they came to be called U-particles — should have a mass he could calculate from the range of the force. The mass of the U-particle should be several hundred times that of the electron.

At that period, physicists devoted much time to the observation of cosmic rays, high-energy particles and gamma rays that impinged on the atmosphere from outer space, mostly originating from outside the solar system and, colliding with one another or with air molecules as they entered the atmosphere, producing showers of secondary particles as expected on the basis of QED. What kinds of particles these were was unknown. Large 'cloud chambers' were constructed in which electrically charged particles would leave visible tracks that were photographed. When subjected to an electric field their paths changed direction, and a magnetic field would curve their paths. From a combination of these effects both their mass and their charge could be deduced. Their charge was negative if their paths curved like electrons, positive if curved the other way.

In 1936 American, British, German, and French teams were puzzled by finding tracks in their cloud chambers that seemed to be caused by particles — they called them 'mesotrons,' — much heavier than electrons but lighter than protons. Could they be Yukawa's U-particles, renamed 'mesons'? However after several years of observations, it became clear that these mesotrons did not interact much with nuclei, as they were able to penetrate deep layers of earth. They could not be Yukawa's mesons. Their mass was measured to be 207 times that of the electron, with a negative charge equal to that of the electron, and they were renamed 'muons.' It took until 1947 for the real mesons to be found, first called pi mesons and then renamed 'pions.' They come in three forms, positive, negative, and neutral, the masses of the charged varieties being 273 times that of the electron and those of the neutral ones 265 times that of the electron.

The cloud chamber had been invented in 1912 by the Scottish physicist C. T. R. Wilson. It consisted of a closed container filled with water and evaporated steam that was at a pressure and temperature to be supersaturated so that it would condense at any slight disturbance,

Figure 9.1. The traces of charged particles in a cloud chamber with a magnetic field. (Source: mrtphysics.co.uk.)

just as clouds in the atmosphere do around dust particles to produce rain. The passage of a charged particle through the chamber would produce tiny droplets from the condensed vapor, causing a visible path along its trail.

This instrument would serve for many years as an invaluable device for detecting charged particles until it was replaced in 1952 by the bubble chamber invented by the American physicist Donald Glaser. The bubble chamber works more or less on the same principle as the cloud chamber, except that it is filled with a super heated transparent liquid such as liquid hydrogen. Any charged particle entering it produces a tiny bubble by causing local boiling and its motion leaves a visible trace. Eventually the unwieldy bubble chamber — both the cloud chamber and the bubble chamber needed to be quite large in order to contain tracks long enough to be observable — was superceded by photographic emulsions and finally by automatic devices: the particles were fed into large arrays of spark chambers, each consisting of two parallel conducting plates with a high voltage difference between them, which produced a spark that was directly recorded in a computer. It was no longer necessary for large numbers of photographs to be examined minutely.

Subsequently particle detectors became both more sensitive and more specialized, growing to become large, heavy machines several of which were needed to surround a collision center to catch all the particles produced. Often such detectors were constructed in distant university machine shops and had to be transported laboriously to the accelerator where they were needed.

Some of the new particles found, however, were unstable, with lifetimes so short that they decayed into others before reaching detectors. To detect such unstable particles, an indirect method is used, based on quantum mechanics. The ordinary particle detectors, located in various directions around the point where the collisions occur, allow the measurement of the total fraction of particles scattered in various directions; this is called the scattering cross section, and it varies with the collision energy. Now, quantum mechanics predicts that the plot of the scattering cross section as a function of the energy will have a noticeable 'resonance' bump at the energy where a new particle is created by the collision, that is, at the energy below which this particle could not be created, so the center of the bump would show its mass. What is more, this bump would be very sharp if the new particle had a long half life and broad if its half life was short: its width should be inversely proportional to its half life. Since most of the many new particles discovered later had very short half lives — otherwise they would have been seen earlier — this is how they were found. As you might expect, that occasionally led to erroneous announcements of discoveries, since a broad natural bump in a cross section plot was hard to be distinguished from a broad resonance bump.

The particle I will discuss in the next chapter, however, is not unstable and hence does not belong to the class to which the above remarks apply.

10

Another New Particle

Another fundamental particle was actually discovered more or less at the same time as the neutron: the positron. This was a particle whose existence and basic poperties had been predicted by an equation in a well-accepted theory: the Dirac equation.

Paul Adrien Dirac, born in 1902 in Bristol, England, was the son of a Swiss-born secondary-school teacher of French and an English mother. He attended the school where his father taught, excelling in mathematics. He had by then developed a passion for math, but in order to avoid a career as a school teacher he went on to get a degree in electrical engineering at the University of Bristol in 1921. Though he was awarded a scholarship to attend St John's College in Cambridge for graduate studies, he could not afford to enroll there without additional support from the local education authority. This needed money was refused because his father had not been a British citizen long enough. He therefore studied mathematics at the University of Bristol, which he was given a chance to attend without paying any fees, and achieved first class honors in 1923. After that Cambridge University awarded him a grant to do research there for his doctoral degree.

Working under the direction of the leading theoretical physicist at Cambridge, Ralph Fowler, who recognized Dirac's unusual abilities, he quickly wrote two papers on statistical mechanics but became fascinated by the then developing quantum theory. Reading proofs of the latest paper by Werner Heisenberg, one of the primal architects of quantum mechanics, he saw a similarity between the structure of Heisenberg's theory and a certain specific formulation of Newton's classical mechanics by William Rowan Hamilton, which led him to construct a fresh architecture for the young theory being invented by Heisenberg and Erwin Schrödinger. These three young men, — well, Schrödinger, born in 1887 was not that young — Heisenberg, Schrödinger, and Dirac (with some assistance from Max Born) thus became the real founders of the fully developed theory of quantum mechanics. For many years Dirac's book, *The Principles of Quantum Mechanics* was the bible for physicists. Though often sought out by other prominent physicists for advice, he had a very strangely withdrawn personality interpreted by some as Asperger syndrome.[1] He had the reputation of saying, very briefly, no more than necessary and exactly what he meant; often just yes or no. Appointed Lucasian Professor of Mathematics at Cambridge, his lectures consisted of reading out loud chapters of his book. However, Dirac continued to make very important contributions and new suggestions in quantum mechanics, quantum field theory as well as cosmology for the rest of his life, and he was much honored by the international physics community. But he was not really happy in Cambridge and retired from the Lucasian chair in 1969, moving with his family to Florida, where he was appointed Professor of Physics at Florida State University in 1971 and continued doing research. Paul Dirac died in 1984 in Tallahassee, Florida, where he was buried.

Returning now to the Dirac equation mentioned earlier, there was one problem with quantum mechanics, and physicists were well aware of it, though for most of their calculations that were compared with experiments it did not matter: the theory, and particularly the Schrödinger equation, was non-relativistic; it did not conform to the special theory of relativity. (Of course, it did not conform to the

[1]For a fascinating biography of Dirac, see G. Farmelo, *The Strangest Man: The Hidden Life of Paul Dirac, Mystic of the Atom*. New York: Basic Books, 2009.

general theory of relativity either, but that did not matter at all until many years later.) Schrödinger tried to make his equation conform to the theory, but when the results were compared with experiments, his attempt was found to be unsuccessful. So Dirac started from scratch and invented a new, relativistic equation for an electron subject to an arbitrary electromagnetic field, quite different from Schrödinger's equation.

The Dirac equation automatically incorporated the correct value of the intrinsic angular momentum of the electron, called its 'spin' (which had to be artificially added to the Schrödinger equation) as well as its resulting magnetism. The value of this 'magnetic moment' was also found to be correct.[2] But more importantly this equation was regarded universally as very beautiful because of its simplicity. Beauty was, in fact, considered by Dirac an important criterion for a good theory, and his equation certainly was a prime example.

The new equation, however, implied a strange prediction: the electron could have an infinite number of negative energies in addition to its observed positive ones. The way Dirac tried to deal with this was to construct an artificial picture in which the universe was full of electrons filling up all the states of negative energy. (Pauli had in the meantime invented the 'exclusion principle' for electrons, which postulated that no available energy level could be occupied by more than two electrons, one with its spin pointing up and the other down. This principle was crucial for explaining the periodic table of the elements by means of the structure of the atom.) However, every so often one of these negative-energy levels was unoccupied, in which case there was a hole that acted like a particle with positive energy and a positive charge. Thus his equation implied the existence of particles with a positive charge of the same magnitude as the electron's! Eventually, when the Dirac equation was reinterpreted

---

[2]Some twenty years later this magnetic moment was experimentally determined to be a tiny bit off, which was eventually extremely precisely explained by the then developed QED.

as a field equation, the artificial hole theory was abandoned, but its prediction of a positvely charged particle (with positive energy) remained. Some physicists at first thought this particle might be the proton, but that did not work: the new particle had to have the same mass as the electron.

The predicted particle, dubbed the 'positron,' was discovered in 1932 by the twenty-seven-year-old post-doc Carl Anderson at Caltech, in his cloud chamber, among the debris caused by cosmic rays. What he saw were some tracks that looked just like those of electrons, except that they curved the other way in the magnetic field. At first thinking they might be electrons coming in the wrong way, he tested this possibility by adding a lead plate that would slow down the particles passing through it, so that he could tell which way they were going — obviously from the fast side to the slow. His observation was soon confirmed by others: he had discovered the particle predicted by the Dirac equation. Evidently these positrons were created in the atmosphere when a gamma ray produced a pair consisting of an electron and its 'antiparticle' a positron. The possibility of such pair creations was predicted by QED.

All subsequently developed quantum field theories would predict, for every particle they described, the existence of an antiparticle with the same mass, opposite quantum numbers and, if electrically charged, the opposite charge. This includes the proton, the neutron, the neutrino, the muon, and the pion. If a particle–antiparticle pair collided, they would both disappear and their energy converted into a gamma ray.[3]

[3]The complete symmetry between particles and antiparticles would later turn out to be not quite exact but only a close approximation.

11

Particle Accelerators

All these discoveries of new particles such as the muon, the pion, and the positron, were made possible by the influx of particles of very high energy from outer space — the cosmic rays. It was clear that for further research, both to explore the structure of the nucleus and to search for other new particles, a more controlled and more intense source of high-energy particles was needed. No heavy charged particles such as protons or alpha particles could overcome the strong repulsion coming from the positively charged nucleus unless they approached it with very high velocity. To accelerate an electrically charged particle to such high speeds, a strong electric field was required, produced between parallel metal plates at high voltage differences. Indeed energies began to be expressed in units of electron volts, eV, the energy to which an electron would be accelerated by the electrostatic field between plates that had a voltage difference of one Volt; eventually, larger units were needed, called MeV (one million eV) and even GeV (one billion eV, the G standing for giga). What is more, since much of the use of these high-energy particles was for the purpose of creating other particles, masses also began to be expressed in the new units; — using $m = E/c^2$ — they are eV/c^2, MeV/c^2 and Gev/c^2. In these units, for example, the mass of the electron is 0.51 MeV/c^2 and that of the proton 938 MeV/c^2.

The first attempts to accomplish such accelerations were made by means of large bulky electrostatic generators invented by the

English physicist J. D. Cockroft and the Irish T. S. Walton. The Cockroft–Walton electrostatic generator was able to produce a voltage difference of 710,000 Volts, enough to accelerate a beam of protons toward a lithium target, transforming lithium atoms into helium atoms, i.e., alpha particles. The American physicist Robert Jemison van de Graaff invented a different kind of generator, the original model of which produced 80,000 Volts but later models were eventually able to generate voltages as high as 5 million Volts. These methods of accelerating particles, however were made obsolete by a totally different kind of machine invented by the American physicist E. O. Lawrence.

The Cyclotron

Instead of accelerating charged particles to their high speeds in one shot by means of an electric field generated by a very large voltage difference, to do it in many small steps, each requiring only a small voltage, was an idea Lawrence picked up by reading the works of the Norwegian physicist Rolf Widerøe, but the ingenious new method of doing it was his own. He bent their paths to be circular by means of a magnetic field. Each semi-circle was housed in a D-shaped metal box, hollow like pita-bread, open at the flat side of the D, the openings facing one another with a gap, and a small voltage difference between them. So when a charged particle crossed the gap from one D to the other at a higher voltage it got a small kick, accelerating it and thereby enlarging the radius of its new path. By the time it then arrived at the gap to cross over to the first D, the voltage difference would be reversed. Now remember that to produce an electric field between metal plates, they have to be at different voltages; so there is no electric field inside a D, and a particle traveling in the metal shell of a D enclosure is unaware of the change of polarity, but when it crosses to the opposing D, there is a field and it gets another kick. As the particle continues circulating through the two Ds, it thus keeps on being accelerated to a higher velocity in small increments. The crucial trick that made all this possible was based on the fact that according to Maxwell's equations, as Lawrence found out, the time it takes for a particle to traverse its semi-circular path through each D, remains exactly the same as the length of its path and its velocity kept on increasing. The needed polarity reversal could therefore be done at a fixed frequency, and new particles fed in bunches into the machine, synchronized with the polarity reversal, would emerge as a beam, all with the same energy.

The first working model he built for himself in 1931 was no bigger than 4.5 inches in diameter; it used a voltage difference of 1,800 Volts and was able to accelerate protons to an energy of 80,000 eV. Soon he built a somewhat larger one, 11 inches in diameter that achieved a proton energy of 1 MeV.

Born in 1901 in Canton, South Dakota, Ernest Orlando Lawrence was the son of parents who had immigrated from Norway. His father was superintendent of schools in Canton. After attending St. Olaf College in Minnesota, Lawrence transferred to the University of South Dakota, where he graduated in 1922. Next year he earned his master's degree in physics at the University of Minnesota. After that he went to the University of Chicago and on to Yale University, where he completed his Ph.D. in physics in 1925 and became an assistant professor in 1927. In 1928 the University of California hired him as an Associate Professor of Physics and promoted him to full Professor two years later, at 27 years of age the youngest Professor at the University of California.

Cyclotrons, based on the invention by Lawrence three years after coming to Berkeley, eventually were built as large as 88 inches in diameter, capable of accelerating protons up to 55 MeV and alpha particles up to 140 MeV. They were very big, heavy machines because of the size of the magnet required to bend the paths of the accelerated particles, they became extremely important instruments for research in nuclear physics. Lawrence's laboratory, called the Radiation Laboratory — Rad Lab for short — in 1936 was made officially a department of the University of California in Berkeley,

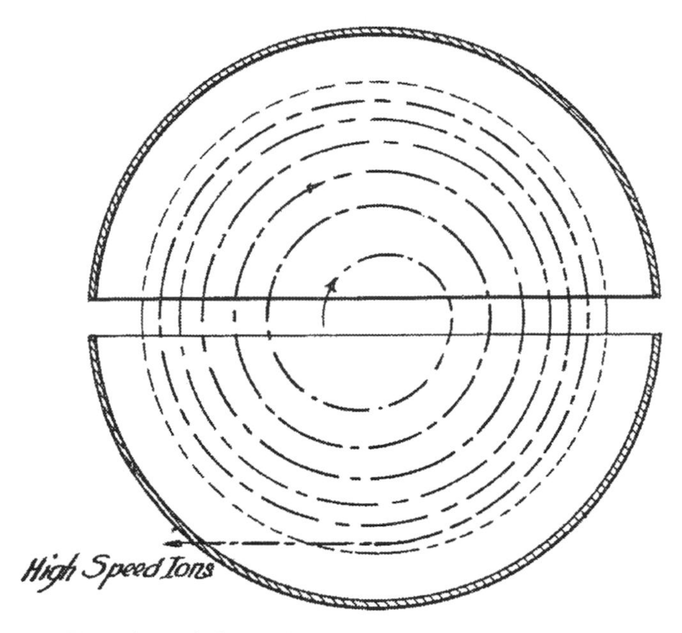

Figure 11.1. Top view of the cyclotron design as shown in Lawrence's patent application. (Courtesy of Berkeley Lab.)

with Lawrence as Director. He remained in that position until his death in Palo Alto in 1958. In his honor, the Rad Lab was renamed the 'Lawrence Radiation Laboratory.'

While nuclear physics remained the forefront of fundamental physics for many years, the time came when 'high-energy physicists' needed particles of higher energies than cyclotrons could provide. The reason was two-fold: Einstein's theory of relativity allowed the kinetic energies of two colliding particles to be converted into mass according to his famous $E = mc^2$, so if there were no other theoretical obstacles like conservation of quantum numbers in the way, new particles could be created, as all the new theories modeled after QED predicted, provided the colliding particles had the needed kinetic energy. The second reason was also based on the theory of relativity, according to which the effective masses of the accelerated particles increased as they neared the speed of light, which gradually changed the needed frequency of the voltage oscillation between the Ds in a cyclotron, making it no longer functioning. New kinds of particle accelerators were required. Indeed, when at the Rad Lab a cyclotron was constructed — with a diameter of 184 inches and a magnet weighing 4000 tons — that would accelerate protons to more than 100 MeV, the limit was reached and the design of the machine had to be altered; it was called a 'synchrocyclotron.'

The Synchrotron and the Big Machines

The main design difference — there were a number of others — between a cyclotron and a synchrocyclotron was that the frequency of the needed voltage reversals was suitably adjusted to accommodate the relativistic effect. The 'isochronous cyclotron,' on the other hand, took the relativistic mass increase of the accelerating particles into account by increasing the magnetic field with the increasing radius of their paths as they moved faster. The later constructed accelerators, which accelerated particles to higher and higher energies — the new theories predicted the possibility of creating particles of ever larger masses — were based on the idea of the synchrotron developed by the Russian physicist Vladimir Veksler in 1944 and the American physicist Edwin McMillan at the Rad Lab in Berkeley in 1945. The first proton synchrotron was designed by the Australian Sir Marcus Oliphant in 1952 and in 1954 such a machine, called the Bevatron, was built at the Rad Lab in Berkeley, accelerating protons to an energy of more than 6 GeV. This energy had been selected specifically to be able to produce antiprotons, the existence of which was confidently expected on the basis of the Dirac equation, by firing protons at a fixed target. Teams led by Emilio Segrè and Owen Chamberlain did find negatively charged particles of the same mass as the proton in 1959. The Bevatron had fulfilled its promise: it demonstrated that the antiproton did indeed exist.

Subsequent particle accelerators, in addition to including great design refinements as well as huge increases in size, incorporated one important innovation. All previous experiments involving particle

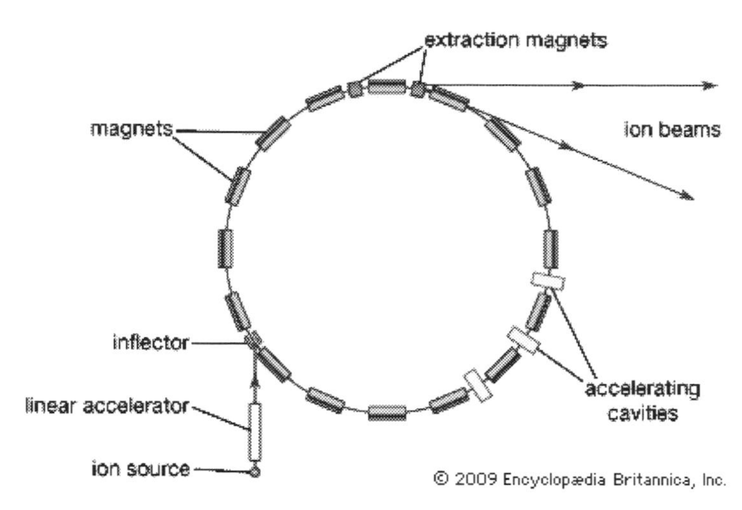

Figure 11.2. One version of design of a synchrotron. (Alternating-gradient synchrotron. Art: Britannica Online for Kids, web, 19 Nov 2013.)

collisions had bombarded fixed targets with high-energy particles. However, the effective collision energy could be much increased if both the bombarding and the target particle moved at high speed against one another, just as the collision of a car moving at 60 miles per hour with an oncoming fast moving truck is more damaging than if the car had collided with a wall. In order to achieve this without having to build two accelerators that shot their emerging particles at one another, the new idea was to have the accelerator inject some of its high-energy particles into a storage ring, where they were kept circulating without further acceleration, and direct the emerging beams from the two against each other. Such a storage ring was a lot cheaper to build than another accelerator, even though it too needed many heavy magnets. Of course, the two intersecting beams would not cause many close encounters or 'collisions' unless they were very well focused. A newly invented beam-focusing method solved that problem, and ever bigger versions of synchrotrons were constructed as new theories predicted the existence of increasingly heavy particles needed to be tested.

Three high-energy particle accelerators were built in the United States. The first, called the Cosmotron, was a proton synchrotron at Brookhaven National Laboratory on Long Island, New York. It reached its full energy of 3.3 GeV in 1953 and continued running until 1968. The second was not a synchrotron but a linear accelerator that became operational in 1966 at Stanford University at the laboratory called SLAC for Stanford Linear Accelerator Center. It is 2 miles long and accelerates electrons to 25 GeV.

For many years the admired and beloved director of SLAC was Wolfgang Panofsky, known as Pief. Born in 1919 in Berlin, Germany, his father a famous art historian, the Jewish family emigrated to the United States in 1934, where his father secured a professorship at Princeton University. Both Pief and his brother Hans, older by one and a half years, immediately enrolled at Princeton and graduated in 1938. The story has it that one of them graduated first in the class and the other second, whereupon they were called the bright Panofsky and the dumb Panofsky (which one was which is still a secret). Pief Panofsky died in 2007 at the age of 88, taking the secret of whether he was, in fact, the dumb Panofsky with him into the grave.

The third high-energy particle accelerator was located at Fermi National Accelerator Laboratory near Chicago, usually called simply Fermilab. It was a proton–antiproton collider called the Tevatron, which began operating in 1983 and was shut down in 2011. The antiprotons were produced by first directing a beam of high-energy protons against a metal target and then accelerated together with the protons in the same tunnel, each beam well focused to keep them separate, but, having opposite charges, going in opposite directions. When the two beams were finally directed to collide, the Tevatron achieved a collision energy of 1.8 TeV. (1 TeV equals 10^{12} eV, i.e., one thousand GeV.)

Finally, there was the Superconducting Super Collider, called SSC for short, to be built in Texas. It was designed to have two colliding proton beams, each of an energy of 20 TeV. Its rings had a circumference of 55 miles. After construction of the needed underground tunnel was well under way and one billion dollars had already been spent, the United States Congress canceled the project as too expensive. Clearly the economic limit of even Big Science had been reached, at least in the United States. Nevertheless, the Large Hadron Collider (LHC) on the border between Switzerland and France was constructed from 1998 to 2008, to accelerate protons to 7 TeV and eventually to 14 TeV. Its circumference is 17 miles, in a tunnel 574 feet below the surface. This large machine was constructed and is run by CERN, 'Conseil Européen pour la Recherche Nucléaire', a fundamental research laboratory located near Geneva, founded in 1954 and supported by twenty European states.

The reason why the rings of such accelerators are so large is that their radius is determined by the strength of the magnetic field needed to bend the paths of charged particles moving near the speed of light. The smaller the radius, the larger the electromagnets have to be, and they are enormous, heavy, and consuming vast amounts of electricity. To save operating costs for electricity, their wires are made superconducting by lowering their temperature to near absolute zero degree K, at which point their electrical resistance is reduced essentially to zero. This is an expensive procedure, but it still saves money.

12

An Abundance of New Particles

Preliminaries

Before describing the large number of new particles that were discovered, some by means of cosmic rays but most by the big accelerators built for that specific purpose, we have to deal with some technical terminology.

The intrinsic angular momentum of the electron, called its spin, was already mentioned in Chapter 10. Quantum mechanics decrees that if a given kind of particle has spin, the only value this spin can have is an integral multiple of $(1/2)\hbar$. ($\hbar$ is the fundamental Planck constant divided by π.) If this multiple is an even number, one simply says the particle has integral spin — which includes the case of zero spin —, and if odd, it has half-integral spin. The electron has half-integral spin, and the photon, integral spin (spin 1).

It turns out that all particles with half-integral spin obey Pauli's exclusion principle (see Chapter 10) and those of integral spin do not. This difference manifests itself, among other things, in the fact that these two classes of particles behave statistically differently when in large numbers. First of all, both kinds, being submicroscopic and thus subject to QM, differ from classically described particles — the statistics for which had been determined in the nineteenth century by Maxwell and Boltzmann — in being fundamentally identical to one another and indistinguishable from others of the same kind; if you detect an electron at one time here and later there, there is no way of telling if it's the same electron (QM prevents you from following its path from here to there).

The following simple example shows why distinguishable and indistinguishable particles obey different statistics. Just count the number of different ways in which two distinguishable particles can be distributed among three boxes: you will find there are nine ways. If they are identical, on the other hand, there are only six ways because (a, b, 0) is the same as (b, a, 0), (a, 0, b) is the same as (b, 0, a), and (0, a, b) the same as (0, b, a). If they obey the exclusion principle, there are even fewer, because (ab, 0, 0), (0, ab, 0), and (0, 0, ab) are forbidden. The kind of statistics obeyed by particles of half-integral spin was developed independently by Enrico Fermi and Paul Dirac and these particles are called 'fermions'; those of integral spin follow a different statistics worked out separately by the Indian physicist Satyendranath Bose and Einstein; such particles are called 'bosons.'[1]

Another thing we should discuss before describing the large number of different kinds of particles that were discovered is the mathematical basis on which sense was finally made of the confusing welter of data and observations.

First of all, there was Noether's theorem, which states in essence that all conservation laws are finally the result of symmetries embedded in the foundation of the equations of motion. Although Emmy Noether proved her theorem in the context of classical physics, its power turned out to be vastly greater in the context of quantum field theory.[2]

The second set of mathematical ideas is the theory of groups, developed in the early nineteenth century by the young French mathematician Evariste Galois for purely algebraic purposes that had

---

[1]This universal connection between spin and statistics is a profound result of relativistic quantum field theory.

[2]For more details on the life and work of Emmy Noether, see Auguste Dick, *Emmy Noether: 1882–1935*. Boston: Birkhäuser, 1981. Also my book, *The Science of Energy*. Singapore: World Scientific Publ. Co., 2012.

nothing at all to do with the important application his theory later found in physics. After his death in a duel at the age of 20, his notes were fortunately rescued by Joseph Liouville, another mathematician.

In the technical sense, a group is any set of mathematical objects, A, B, C,... for which multiplication is defined, so that C = AB and D = BA are also members of the group. (AB need not be equal to BA, i.e., multiplication need not be commutative; if it is so for all pairs of group elements, the group is called Abelian.) Furthermore there is a unit element E, so that for every A in the group, AE = EA = A, and every member has an inverse in the group, so that $AA^{-1} = A^{-1}A = E$.

Here is an example: the group S_3 of permutations of three objects. Calling $E = (123)$, $A = (132)$, $B = (213)$, $C = (321)$, $D = (231)$, $F = (312)$, where $(132)[abc] = [acb]$, etc., and multiplication is defined by the following table, which lists the product of each element of the first row by every element of the first column. For example, to find the product BC, look at the entry in the row labelled B and the column labelled C, where you will find D: so $BC = D$.

E	*A*	*B*	*C*	*D*	*F*
A	*E*	*D*	*F*	*B*	*C*
B	*F*	*E*	*D*	*C*	*A*
C	*D*	*F*	*E*	*A*	*B*
D	*C*	*A*	*B*	*F*	*E*
F	*B*	*C*	*A*	*E*	*D*

The members of the group may be general operations, such as translations, reflections, or rotations. (As you can easily verify, the group of rotations of a sphere is not commutative.) Rotations by arbitrary angles are, in fact, an example of a continuous group, the theory of which was founded in the later nineteenth century by the Norwegian mathematician Sophus Lee and further developed by the French mathematician Élie Cartan.

A representation of a group is a mapping of every group element onto a square matrix (we will be dealing with square matrices only, i.e., matrices with as many rows as columns) in such a way that group multiplication corresponds to matrix multiplication. (Matrices are arrays of numbers that used to be quite unfamiliar to physicists. When Heisenberg invented quantum mechanics, he arranged experimental data in arrays and made them obey multiplication rules that turned out, strangely, to be such that AB was not equal to BA. It was Max Born, whose assistant Heisenberg had been, who had to explain to him that these arrays were nothing but matrices, objects whose algebra mathematicians were quite familiar with.)

Matrix multiplication is defined as seen in the following examples of 2×2 matrices

$$\begin{pmatrix} a & b \\ c & d \end{pmatrix} \begin{pmatrix} e & f \\ g & h \end{pmatrix} = \begin{pmatrix} ae + bg & af + bh \\ ce + dg & cf + dh \end{pmatrix}$$

and 3×3 matrices:

$$\begin{pmatrix} a & b & c \\ d & e & f \\ g & h & i \end{pmatrix} \begin{pmatrix} j & k & l \\ m & n & o \\ p & q & r \end{pmatrix}$$

$$= \begin{pmatrix} aj + bm + cp & ak + bn + cq & al + bo + cr \\ dj + em + fp & dk + en + fq & dl + eo + fr \\ gj + hm + ip & gk + hn + iq & gl + ho + ir \end{pmatrix}$$

Now here is the purpose of using such an esoteric mathematical device as group theory. If the underlying theory with its equations is invariant under the operations of the members of a certain group, then it is possible to ascertain, for each energy level, the number of states of the system that have the same energy, without having to solve the very complicated equations. In the context of quantum field theory, 'energy level' of course means the energy — and hence mass by $E = mc^2$ — of any new particle predicted to be produced in collisions of sufficient energy. If the mass of the new particle belongs to an n-dimensional representation (i.e., whose matrices are $n \times n$) of the relevant symmetry group, that means that this particle comes in n different forms, with different properties, such as electric charges, for example, or half lives, but all with the same mass. (Note that a given group may have several different matrix representations of the same dimensionality.) So very important facts about the new particles predicted by the theory can be established on the basis of its symmetry group without having to solve any of the field equations, which usually is extremely difficult and time-consuming. The nature of the representations to which specific particles belong also provides information about the transition probabilities from one to another, if unstable, and which of these transitions, if any, are strictly forbidden by the theory.

The New Particles

The first new particles, discovered both in cosmic ray showers as well as in the beams of some of the big particle accelerators, were unstable and considered strange. Strange because, on the one hand, they were relatively easy to produce, i.e., more of them appeared as the result of collisions than were expected, but, on the other hand, their half lives were relatively long, so that their trails were easily visible in bubble chambers and photographic emulsions. On the basis of quantum field theory one would expect essentially the same kind of interaction to be responsible for the creation of a particle as for its eventual decay into others, so copious production — indicating strong interactions — should go with a short half life.

The solution to the riddle of strangeness was solved relatively easily by postulating that these strong interactions obeyed a new conservation law. There was a new quantity — it was naturally called 'strangeness,' which could be positive or negative and it might be a quantum number rather than a continuous quantity — that was conserved in all interactions. Thus strange particles would always have to be produced in pairs with compensating amounts of strangeness, one having as much positive strangeness as the other had negative. However, their decay was forbidden because it would violate the conservation of strangeness. The only reason why they would eventually decay nevertheless was that there was another, weak interaction at work that violated the strangeness conservation law.

So let us look at the first new particles that were found. There were two classes of them: the first — named 'hyperons' — consisted of fermions heavier than nucleons called *lambda* (neutral), *sigma* (positive, negative, and neutral), and *xi* (negative and neutral); the second class was made up of bosons lighter than protons but heavier than pions — the members of this class were named K-mesons or 'kaons' (positive, negative, and two different kinds of neutral ones). In order to account for them, the American physicist Murray Gell-Mann and the Japanese Kazuhiko Nishijima devised a scheme to define the strangeness in such a way that there ought to be two kinds of neutral kaons with different half lives. This prediction was soon verified by experiments at the Cosmotron accelerator at Brookhaven National Laboratory.

Many of the particles discovered, however, did not have long enough half lives to leave visible tracks in bubble chambers, cloud chambers, or photographic emulsions. Their existence had do be inferred from resonance bumps in scattering cross section plots as explained earlier. The big question remained by what scheme to explain, for example, the existence and lifetimes of the newly discovered hyperons. What was the underlying symmetry that accounted for the essentially equal-energy quantum states — the almost equal masses of the corresponding eight particles that came to be called 'baryons' — the six hyperons and the proton and the neutron? The responsible symmetry group, called SU(3), was found independently by the Israeli physicist Yuval Ne'eman and by Murray Gell-Mann. The group SU(3) consists of transformations in three dimensions — though these are not in physical space but in an abstract mathematical one — had indeed an eight-dimensional representation, called by

Gell-Mann 'the eightfold way' that accommodated exactly the eight baryons. Not only that, but the same symmetry group also accounted for the seven known mesons, i.e., the three pions and the four kaons, plus the subsequently discovered meson called *eta*.

The only slight flaw in this scheme was the fact that if the underlying (unknown) equations were indeed symmetric under the SU(3) group, then the eight baryons should all have the same mass, and so should the eight mesons. But experiments showed that these masses, though close, were not exactly equal. Gell-Mann and the Japanese-born American physicist Susumu Okubo thereupon invented a way in which the SU(3) symmetry was not quite exact but slightly broken: the symmetry breaking would lead to small changes in particle masses, and results of numerical calculations agreed reasonably well with the experimental data.

Meanwhile nine new particles had been discovered that appeared to be excited states of the baryons. Four seemed to be excited states of the two nucleons: neutral, negative, positive, and doubly positive; these four were named *delta*. Then there seemed to be one excited state for each of the three *sigma*s and of the two *xi*s. These nine states seemed to fit exactly into a ten-dimensional representation of SU(3), with all the correct charges. Except that the tenth place, which should be occupied by a negatively charged particle of known quantum numbers, already named the *omega-minus*, was empty. Using the Gell-Mann–Okubo mass formula, even the mass of the missing particle was approximately known. Finally, two years later, the predicted *omega-minus* was found in a bubble-chamber photograph at Brookhaven National Laboratory. The fact that this missing tenth particle was discovered after the promulgation of the theory, where it filled an empty hole, was especially convincing evidence for the correctness of the scheme, just as much earlier had been the later filling of the initial gaps in Mendeleev's periodic table of the elements.

The problem of the strong interactions, those responsible for the stability of atomic nuclei (absent beta decay), remained to be understood. Having succeeded well with SU(3), Gell-Mann proposed to use what is called the 'fundamental representation,' which is three-dimensional. All particles regarded as elementary up to now, he postulated, were really made up of more basic fermions which he named *quarks*, picked from a line in James Joyce's *Finnigans Wake*. Each of the three quarks would have to have a fraction of the charge of the electron, one of them negative with 1/3 the electron's charge, and two positive with 2/3. The Russian-born American physicist George Zweig had come up independently with a very similar scheme, naming the new particles 'aces', but 'quarks' was universally adopted.

As far as Gell-Mann was concerned, the whole idea seemed to be just a mathematical scheme devoid of reality: each baryon consists of three quarks, each meson (including some newly discovered ones, all of them bosons) of a quark and an antiquark. Nevertheless, it did explain why it was that nature utilized only the eight-dimensional and the ten-dimensional representations of SU(3). In spite of long experimental searches, no particles belonging to other representations of the same group were ever found. What is more, new scattering experiments seemed to indicate that nucleons appeared to be made up of 'partons,' somewhat analogous to the way the Geiger–Marsden experiments had convinced Rutherford that the atom contained particles inside. So quarks were more than a mathematical fiction but did seem to be real objects, despite the fact that to this day, no particle of fractional electronic charge has ever been discovered outside the confines of baryons and mesons. But there is no question that this counter-intuitive notion of the most fundamental building blocks of nature having fractional electric charges brought order to the disconcerting welter of 'elementary' particles that experiments had shown to exist. Even more or less plausible ingenious theories were invented to explain why quarks could never be found isolated in the wild but only inside the cages of certain other particles. To go into these theories, though, would lead beyond the purpose of this book.

There was, however, another problem with quarks as the fundamental building blocks of baryons and mesons: they had to have half-integral spin like electrons and protons and hence they had to be fermions, obeying the exclusion principle. That, however, did not fit the experimental data. The solution of this conundrum was proposed by the Japanese-born American physicist Yoishiro Nambu and the Korean-born American Moo-Young Han. In order to make it possible for two quarks to occupy the same state, they introduced a new quantum number named 'color,' even though it had really nothing at all to do with the usual meaning of that word; it was entirely metaphorical: each quark, they postulated, came in three different 'colors.' The result was, of course, that two quarks of the same traditional quantum numbers but different 'colors' could occupy the same state, and that brought agreement with the experimental data. From then on, 'color' was a quantum number that had to be assigned to other particles as well.

This was not the end of the necessity to introduce more quarks with new quantum numbers. One was the parallelism between the classification of the hadrons (strongly interacting particles) and that of the leptons (electrons, muons, and neutrinos), a parallelism required in order for the theory to be free of infinities (renormalizable). But there was one quark missing: there were four leptons, but only three quarks. This required the introduction of another quark that had the distinction of being *charmed.* When the American experimental physicist Martin Perl discovered another lepton, more than twice as heavy as the proton, which was named *tau,* the number of quarks again had to be increased, finally to six. (The tau plus its associated neutrino increased the number of leptons to six.) They were distinguished by having six different 'flavors' named *Up, Down, Strange, Charmed, Top,* and *Bottom.* (The last two were also sometimes called *Truth* and *Beauty.*) Gone is the traditional scientific custom of choosing Latin or Greek names for newly discovered phenomena. It took several years for the conjectured (or theorized) new particles to be experimentally found, but found they were. And there can be no question that the group-theory based ordering of the confusing discoveries of ever-new and more particles simplified our basic understanding as much as the structure of the atom discovered by Bohr and Rutherford simplified chemistry by furnishing the basis of the periodic table of the elements.

There were three more particles, theoretically predicted but waiting to be discovered. The first two had to be introduced to explain the long-known phenomenon of beta-decay. The theory that Fermi first invented turned out to be not feasible. So the Indian-born American physicist E. C. G. Sudarshan and the American Robert E. Marshak constructed a new theory modelled after QED, which required them to postulate two new particles called W and Z. It took, however, more than twenty years before these two were finally discovered by the Italian physicist Carlo Rubbia and the Dutch physicist Simon van der Meer at the CERN laboratory.

This takes us almost to the end of the list of particles. There was, however, still a capstone missing. All the new field theories constructed more or less after the model of the highly successful QED had one characteristic in common: they were what was technically called 'gauge theories,' the exact meaning of which need not concern us here, but which had the crucial property of being 'renormalizable,' i.e., free of nonsensical infinities. Because it was a gauge theory, QED, the quantum field theory of electromagnetism, produced — in a metaphorical sense — the particles of light known as photons, which were massless, i.e., their 'rest mass' was zero. Similarly, so did all other gauge theories: none of the particles they 'produced' had any mass — rest mass, that is. So why did all the many existing particles described above have any mass? The theory that finally managed to have the effect of providing all the other particles with mass — except, of course, the photon — was surprisingly based on a variety of ideas imported by Philip Anderson and Yoichiro Nambu from the theory of superconductivity as well as ideas by the English physicists Jeffrey Goldstone and Peter Higgs. This crucial theory, responsible, in a sense, for all the mass in the universe, 'produced' its own heavy particle that became known as the Higgs boson. As its expected mass was only vaguely known, it took more than 40 years betweeen its first prediction by Peter Higgs and its experimental discovery this year (2013) at the Large Hadron Collider, or LHC at CERN, an accelerator constructed pretty much for the very purpose of finding the elusive Higgs boson. The Tevatron at Fermilab, built mostly for the same purpose, was shut down in 2011.

The theoretical expectation at this point in time is that we now know all the fundamental particles that exist in the universe. The theory known as the *Standard model*, which is believed to be the rock bottom quantum field theory, does not predict any others. But since the biggest problem in theoretical physics, the reconciliation of quantum theory with general relativity is still unsolved, who knows what the physics of the twenty second century will bring?

References

Dick, A., *Emmy Noether: 1882–1935*. Boston: Birkhäuser, 1981.

Farmelo, G., *The Strangest Man: The Hidden Life of Paul Dirac, Mystic of the Atom*. New York: Basic Books, 2009.

Newton, R. G., *The Science of Energy*. Singapore: World Scientific Publ. Co., 2012.

Pais, A., *Niels Bohr's Times in Physics, Philosophy, and Polity*. Oxford: Clarendon Press, 1991.

Index

Printed in the United States
By Bookmasters